COST IMPROVEMENT,
WORK SAMPLING,
AND
SHORT INTERVAL
SCHEDULING

COST IMPROVEMENT, WORK SAMPLING, AND SHORT INTERVAL SCHEDULING

WALLACE J. RICHARDSON
Professor of Industrial Engineering
Lehigh University

RESTON PUBLISHING COMPANY, INC.
Reston, Virginia
A Prentice-Hall Company

Library of Congress Cataloging in Publication Data

Richardson, Wallace J
    Cost improvement, work sampling, and short interval
scheduling.

      Includes index.
      1. Costs, Industrial. 2. Work sampling. I. Title.
HD47.R46   658.1'552   76-7626
ISBN 0-87909-139-8

© **1976 by**
Reston Publishing Company, Inc.
A Prentice-Hall Company
Reston, Virginia 22090

10   9   8   7   6   5   4   3   2
Printed in the United States of America.

To my wife

CONTENTS

PREFACE

This book has two objectives. First, it is intended to serve as a guide for managers and industrial engineers in the institution and operation of cost improvement programs. Secondly, the book presents what is intended to be a brief but useful discussion and update of the techniques of work sampling and short interval scheduling. In broader terms, it is hoped that this book will be useful to those who are concerned with the improvement of the cost of operations of all sorts manufacturing, service and government institutions—by providing a definitive treatment of cost improvement programs, and a "how to" reference for the techniques described. The book is not intended to be used primarily as a college textbook. Rather, the approach is directed to the practical needs of the operating manager and his staff.

Since there are two objectives, it seems sensible to discuss them in turn. "Cost improvement" is sometimes termed "productivity improvement," "cost reduction" or "profit improvement." There are semantic differences here, and these differences can be significant, but the basic concept is well understood. Simply stated, most managers realize that any improvement (reduction) in their present cost of operations will have an immediate and positive effect in a number of directions. The problem has been how to accomplish such a change in a systematic and effective manner. It is the contention of the author that an organized cost improvement program offers the manager a vehicle which enables him to involve his entire organization in a constructive effort toward improvement, which enables him to utilize his previously developed skills as a manager, and which can be presented fairly as a means of accomplishing an important organizational objective without diminishing the self-esteem and individuality of the members of the organization. Further, it has been the author's experience that without such an organized approach, cost improvement activity tends to be spotty, ineffectual, and indeed often results in serious personal and organizational conflict. The first section, therefore, presents a discussion of the salient features of successful cost improvement programs as well as specific suggestions which should be helpful in making such a program work.

It should be pointed out that the approach to cost improvement programs given by the author is generally limited to the improvement of existing operations. This in itself is a rather ambitious objective, although it deals with a matter of primary concern to managers. But it should be realized that areas such as new product development, product and process research and the restructuring of corporate funding—all potential sources of increased profit—are not dealt with. No one book can do justice to all courses of cost improvement. This book concentrates on the improvement of day-to-day operating costs in the areas usually under the control of the top local organizational managers. It is hoped that this fills a compelling need, whether that organization be a manufacturing plant, an office, a hospital, a store, a utility or a government agency. No one has unlimited resources; all are subject to some economic constraints. Systematic improvement of operating costs is therefore vital to continuing success and even to survival. Since no one seriously anticipates lower wage rates, lower basic material costs or lower taxes, the problem is to use these resources prudently to produce the desired output. This book suggests a practical approach to meet this objective.

The second part of the book consists of discussions of work sampling and short interval scheduling. Neither of these is a new technique. Work sampling in particular has many uses beyond its integration into cost improvement programs. But work sampling is an analysis technique which is fundamental to the appraisal of the activity of people or machines—which may be substantial elements of operating costs. The presentation of work sampling is not really definitive, but is intended to be a step-by-step set of instructions, which reflects current practice in the technique. Short interval scheduling is, again, not new. But it is used more and more as a tool for supervisory development and as a means of improving productivity in indirect-type work. The technique is dealt with here as a practical, do-it-yourself application of good management practice. Short interval scheduling is basically a straightforward procedure, well within the capabilities of typical operating management.

This is a book for the manager and the practitioner. The material in this book reflects the experience of the author, of course. However, this experience was gained by working with many people in many organizations. Space simply does not permit the acknowledgement by name of all of the people to whom the author is indebted. There are, however, some who were particularly helpful in the development of this book. First, thanks are due Lehigh University for providing an environment which allowed me the freedom to develop ideas and to test these ideas outside the classroom. I am grateful to Professor William T. Morris of Ohio State University for his perceptive and constructive review of the original manuscript. And I am indebted to Professor George E. Kane, chairman of the Department of Industrial Engineering at Lehigh for his unfailing encouragement and superior professional advice. It is my hope that this book will do justice to those who have helped so much in its development.

WALLACE J. RICHARDSON, P.E.
Bethlehem, Pennsylvania

1

CHARACTERISTICS OF COST IMPROVEMENT PROGRAMS

Our objective in Part One is to advance the viewpoint that a systematic program of cost improvement is essential to the realization of the aims of management and to describe the characteristics that are generally found in such programs. Not every successful cost improvement program will have all these characteristics, of course, because the individuals, organizational structure, objectives, and past histories of the companies and institutions concerned are so different. But a common thread through all successful efforts at cost improvement seems to be an organized program; many or all of the characteristics described here are commonly found in such programs.

Techniques of analysis and improvement will not be dealt with exhaustively in Part One. Rather, the establishment of the program, the demands upon the personnel involved, and, in general, the features of a manageable program will be examined. Techniques will be discussed in Part Two. As a general observation, most techniques of analysis and improvement are not new. The flow chart and work sampling have been around for quite a while, for example. But to be really effective, these techniques must be applied within a framework that sets objectives for their use, plans and schedules their application, allows for their proper application, and provides systematically for some sort of action and follow-up. This framework, which is recognized as the management process, is the subject of Part One.

Need for Systematic Formal Cost Improvement Programs

Cost improvement is generally recognized as an extremely desirable type of activity, which should improve the ability of a company to compete in its field or an institution to justify its level of expenditure. It is a superb vehicle for the development of managers and supervisors at all levels of organization. It tends to stimulate everyone involved to look for new ideas and to question current practices. Cost improvement almost always brings with it some element of conflict, however. For example, high inventory levels help improve customer service and manufacturing scheduling. Yet to carry inventory costs money, and any "improvement" in the form of reduced inventory will be met with resistance based on the difficulties forecast in these operating areas. Management's job is to see that the proper balance is reached. Cost improvement programs should improve our operating effectiveness and the utilization of the resources at our disposal. Yet it is easier to work with surplus people and loose budgets. The resolution of this conflict is what management is all about.

EXECUTIVE RESPONSIBILITY

Managing a business to compete in its field and produce a profit, or managing an institution to provide health, education, or public services, may be compared to a juggler's balancing act. The chief executive must constantly reconcile the legitimate demands for attention from the sales manager, the controller, and the manufacturing superintendent. He must keep in perspective the requirements of good customer service, sound financial control, and production efficiency. He must realize that there will inevitably be conflicts, and he alone has the responsibility of resolving these in the interest of the company as a whole. The chief executive cannot afford to become too involved in any particular aspect of the business as a continuing effort, because situations change almost overnight. The quality problem that absorbed his attention this week may be replaced by a labor relations problem which becomes more imperative next week. The executive's dilemma is that he is responsible and accountable for all areas of the operation, and yet he must be free to devote his time to particular areas as compelling problems arise. To recognize these problems early, and at the same time to maintain control across the board, he usually institutes management systems and programs that will accomplish through others the day-by-day operations leading to the achievement of objectives. This is, in Peter Drucker's phrase, "the art of management." This is what the good manager does best. But the manager must have something tangible to manage. He is most effective when he is involved in the classical sequence of setting objectives, planning, scheduling, execution, and follow-up, which characterizes the process of management. Our interest is in cost improvement; to achieve this, there must be a formal, systematic program that will allow management to function in its normal pattern.

FORMAL COST IMPROVEMENT PROGRAMS

The Need For Change

One basic strength of the management process is its ability to handle change. Both cost improvement and the competitive environment demand change, and it is natural for management to define the goals inherent in these changes and to absorb them into delegated responsibilities. The need for change may become apparent through the systematic feedback control of information systems, or it may become apparent through something much more dramatic. Recent examples of the latter abound, as in the fluctuations of the money market, the growth of worldwide shortages of some vital materials, the sharp increase in the cost of energy, and the growth of government involvement in environmental aspects of plant operation, safety, and consumer protection. There is no problem in getting management's attention in such cases. Indeed, the drastic effect of these phenomena demands attention. Such demands seem to run in cycles, however, and each crisis has a way of being settled and reduced to operating company policy. It is a truism, though, that many company policies set under such duress seem to persist long after the crisis has been met, and need some systematic review. Many companies, for example, find that they have failed to readjust inventory levels from a high point caused at some time in the past by a critical shortage, which is no longer a problem.

Attitudes Toward Change

In addition to the dramatic circumstances that almost demand change, there are many areas where change has not taken place, but which should be reviewed to see if change should have occurred. The adage, "no news is good news," may not be true, for it may reflect a situation in which no strong effort is being made to take advantage of technological improvement or to reflect the natural learning process that should result in cost improvement. Supervisors in such areas may be doing a good job of administration, and will probably feel put upon when the pressure for improvement begins to be felt. After all, they may feel, they are doing a good job now so why change? And while it is self-evident that to improve we must change, it also is true that change does not necessarily bring improvement.

Formal cost improvement programs offer management an opportunity to bring into question the existing pattern of operations in all parts of the business. In a sense, they put pressure on individual managers to review current practice. Everyone has a very human tendency to defend his own past performance. And implicit in any suggestion for change is criticism of the way things are done now. So the manager who is, in fact, doing a good job now may feel that it is unreasonable to expect him to seek change—and improvement is change. Yet if each will recall the past, he will realize that one thing that has made him successful has been an ability to adapt to new techniques and demands. This may be true even if he did not initiate the changes. The manager's job is to keep his area of responsibility competitive, and he must provide for the systematic planned improvement necessary to do this. Cost improvement programs of a formal nature help meet this demand.

The Learning Curve

Frequently, a supervisor who is doing a capable job will point out that he has the reputation of running an efficient department, and that he has been successful over the years in meeting cost and performance targets. Because things are running smoothly, he may believe that he has more limited opportunities for cost improvement than do other departments where problems are acknowledged to exist. This may be a fair statement, but not necessarily a complete one. Performance over time of substantially the same task should result in improvement. In brief, learning should take place and with it cost improvement. This phenomenon is widely recognized, accepted, and formalized as a learning curve or progress function. The *learning curve*, simply stated, takes the form of a progressively lower time or cost as successive units are produced or procedures repeated. It is reverse-exponential in form; that is, the second unit requires, for example, nine tenths of the time or cost of the first; the fourth, nine tenths of the second; the eighth, nine tenths of the fourth, and so on. Thus, even if an operation has been standardized for some time, there should be some cost improvement inherent through repetition alone. The learning curve, also known as the manufacturing progress function, has been shown to apply not only to the manufacture of aircraft, appliances, electronic products, and short runs in job-shop manufacturing, but also to labor standards and even to the floor space required to produce a specific product. These last two items may seem unusual, but we have only to observe the "loosening up" of rate structures and the seemingly elastic properties of factory floor space to realize that the phenomenon is real. So, at the very least, we should expect from stable operations cost improvement due to the learning effect.

Rationale For a Formal Program

The basic rationale then is that cost improvement is important, that a formal program helps to make this cost improvement manageable, and that a fundamental of the program is that eventually all parts of the operation will be covered. If top management subscribes to this, it will find that the constructive pressure for cost improvement will bring out in the open many fundamental questions which otherwise may not become a matter of serious discussion until a crisis arises. Typical questions are: 'What should our customer service policy be?', 'What is our plant capacity under current conditions?', 'What should our plant capacity be?', 'Does our policy toward overtime lead to higher costs?', and 'Do the inventory restrictions imposed two years ago make sense today?' Such questions deserve thoughtful answers, but sometimes the questions themselves are never asked because historical goals are being met, and no one wants to "rock the boat."

The last paragraph raises the point that department heads often regard themselves less in competition with other companies than with other department heads. Cost improvement programs provide an orderly means for top management to put in perspective for everyone in the company the ugly fact that the customer does not distinguish between a vendor's sales force and a vendor's production department. The customer views the vendor as one unit source, in competition with other unit sources. It is the executive's job to manage the company as a competitive entity, and he needs every device he can employ to do this. Cost improvement programs are most effective in bringing about an appreciation of the company's need to be competitive in all areas. And cost improvement programs also bring pressure at the points where the need is greatest, that is, where the separate functions which expend resources to produce product or service are managed.

As an extension of the previous discussion, formal cost improvement programs operate through the achievement of a series of specific *short-term goals*. More will be said later about the process of setting these goals and auditing performance against them. It is necessary to the management process that cost improvement programs include the assigning of responsibility and accountability for performance. This process is understood and accepted in many other parts of the organization. The sales function has forecasts and quotas, engineering has target dates and costs, the controller has financial limits and ratios, and manufacturing has standards of cost and quality to meet. Managers at all levels are accustomed to the process of setting goals, planning, scheduling, execution, and follow-up. Indeed, they expect to work in such an environment, and they are right. A manager simply cannot function well in directing the efforts of his people if he does not have specific, short-term goals. This is almost self-evident, but too often cost improvement takes the form of inspirational messages and arbitrary budget cuts, neither of which seem to result in continuing improvement.

The setting of short-term goals will of course stimulate discussion of the real long-term objectives to be sought. This is important, to be sure, but long-term objectives tend to be general in nature, because they must encompass a variety of functions and because so many changes occur over time in the industry and in the marketplace. And long-term objectives, no matter how clear, still must be translated into specifics for each department. The short-term goals typical of cost improvement, furthermore, serve as a focus of decision. In the process of formulating and reviewing these goals, the executive has an opportunity to educate his group of managers; at the same time he receives feedback that he may not get in any other way. For when a manager is faced with the prospect of making a firm commitment to his boss, he will be much more realistic in describing situations as they really exist. This realism is most valuable, and it seems to be elicited most effectively when there is pressure to meet specific objectives.

The discussion of specific short-term goals is also the point at which subordinates will test the executive's basic convictions, and his willingness to make the hard decisions that will be necessary to achieve real cost improvement. The executive should be ready for this. For example, a production supervisor may say that his ability to meet his cost improvement goals depends largely upon strict observance of a rule (already agreed upon and made company policy) that limits schedule changes to meet certain types of delivery promises made by sales. He can reasonably ask that the top management either enforce the rule, change it, or not hold him responsible for cost overruns due to nonenforcement. This is the type of situation that must be resolved at any rate, but the cost improvement program tends to bring out the points at which executive decision is necessary. An advantage of this chain of events is that, once the question is resolved, pressure to make the solution work falls on those who raised it.

An additional and substantial benefit to management of the introduction of a formal cost improvement program is that the experience of working within such a program is invaluable to the personal development of the individuals involved. Not only do they receive reenforcement in the use of the basic management process, but they also perceive on a broad scale the interrelationships of functions within the company. Everyone given responsibilities within the program is put in a position where cooperation with others is a necessity. At the very least, there should emerge a universal understanding of the accounting system and of the importance of sound internal financial control. Finally, a part of most cost improvement programs is the exposure of those participating to new techniques and new ideas. This can be done in other ways, but this way seems more real.

With all the advantages of introducing formal cost improvement programs, one fact must be recognized: although cost improvement may be said to be an almost universal and continuing need, it is by no means always the most important. The really vital areas of interest for management may lie in new product development, marketing, or the company response to a crisis outside its control. Management must be free to concentrate on these as they arise. Furthermore, many highly profitable companies depend on free-wheeling innovation rather than efficient operation for their success. But any cost improvement effort, when done reasonably and well, cannot help but improve a company's ability to compete successfully.

In some operations, such as health care, education, and government, where the product is more difficult to evaluate and where there exist external influences that sometimes override internal management options, cost improvement still should be sought in an energetic fashion. It is becoming more and more apparent that these areas face the problem of allocation of limited resources, rather than the problem of establishing the basic desirability of the function itself. It is self-evident that we need good hospitals and patient care, good schools and instruction, and good government services. But in competing for funds, sometimes from a single source, to meet all these objectives, a case must be made that these funds will be spent prudently and that fair value will be received. It is becoming obvious to administrators in institutions that cost improvement programs are just as important to them as they are to executives of profit-oriented businesses.

AN ALTERNATIVE

To complete our discussion, an alternative to our point of view should be presented. Assuming that cost improvement itself is seen to be desirable, the most common way for managements to achieve this has been simply to cut the input, that is, reduce the funds available in the budget. Budget cutting is usually done on a flat percentage basis, and usually indirect expense is cut the most. The chain of events follows a pattern: a periodic financial report shows reduced earnings, the executive goes over his own budget analysis, and realizes that he really has no solid input-output information, and so he decides that the fairest way to improve costs is to cut across the board. This type of indiscriminate slash is known as the "meat-axe" approach or "management by fiat." Despite the apparently arbitrary nature of this approach, it has both advantages and disadvantages. The advantages of the flat budget cut include the following:

1 / It is simple. The executive does not have to concern himself with staff support or long-term implications of a policy nature. He just writes the letter and sends it to everyone.

2 / It works. It may work only in the short term, but when a subordinate receives a letter that directs him to cut the cost, regardless, he cuts the cost, regardless. It can be argued that this approach is like that of deferring maintenance in that the long-term effects will be more costly than the short-term savings. But the executive is more concerned about an upcoming board meeting and low earnings than he is with the long term. After all, if he cannot improve costs, he won't be around in the long term.

3 / If the program works, the executive is the hero. It is his name on the bottom of the directive.

4 / To be objective, such a flat cut, particularly in indirect areas, may be a good thing. Any department that has not been working against systematic goals for improvement probably has some slack. Every experienced manager knows that working against standards or goals improves efficiency, and so the first broad cut may simply cut out this slack. To be blunt about it, the first cut usually gets only the fat. (The problem then is what following steps should be taken.)

These are attractive advantages, and the executive may regard them as compelling enough, even though he knows that he may be building trouble in the future.

The disadvantages of the flat-cut approach include the following:

1 / The supervisors who are now doing the best job are penalized most. The efficient suffer more by a flat cut, because they have less fat in the budget.

2 / When business is bad (the usual reason for a flat-cut letter) there may be uneven effects on operations, and yet the cut usually is made across the board.

3 / The "meat-axe" approach is difficult to justify as a realistic management pattern. It sets a precedent that may return to haunt the boss.

4 / The executive can be certain that once he sends out his first "meat-axe" letter he has already received his last realistic budget request from most of his subordinates. The obvious course for them in response to what they recognize as an arbitrary procedure in setting goals will be to inflate their budget requests. Management then becomes a matter of strategy and gamesmanship, which is not unheard of. But it is no way to run a business.

These difficulties have proved to be real. But we should not overlook the fact that the flat-cut approach has advantages, particularly the first time it is used. In fairness to the executive, the "meat-axe" is a highly visible response to urgent needs for cost improvement. It is the thesis of this work that a formal cost improvement program is a better way, but the program must be presented to top management in a positive fashion. That, too, is a thesis of this book.

In summary, the advantage to management of a formal cost improvement program is that the following objectives can be met:

1 / Giving management something to manage; that is, a process within which they can exercise proved skill.
2 / Exerting systematic pressure for improvement as part of the regularly accepted responsibilities of supervisors.
3 / Providing an opportunity for personal development.
4 / Serving as a channel of communication (both upward and downward) within the company.
5 / Most important to all, usually reducing costs and improving the company's ability to compete in the marketplace.

Like all desirable objectives, however, these things are not easy to do. But managements have not much leeway in operations now, and now is the time to start taking positive action toward cost improvement. It will not come from the lower echelons of the organization. It is top management's job.

This section has discussed the responsibilities of management, and the need for management direction. Any time there is a change, there is also an implied criticism of the current situation. Management probably is largely responsible for this situation. In most cases, management neglect of ongoing cost improvement and the lack of management perception toward the development of rising costs have been principal contributing factors to the need for a cost improvement program. This may be an unpalatable truth, but, nevertheless, it usually is a truth. It is suggested that management face this, and make some point in its conduct of the program that management, too, realizes some responsibility and that management practices will be made the subject of the same sort of critical review that management is asking of the entire organization.

The form which the management effort will take varies. Part of this effort will come in a critical review of administrative expenses. Part will come from a deliberate effort to develop management skills. But the overwhelming contribution which management can make is for each person in management to be honest with himself about the areas of his responsibility which need improvement, and to communicate to those of whom he is asking greater effort about his own dedication toward improvement. The manager will find, incidentally, that if he combines this attitude with the deliberate development of better skills in interpersonal relationships, the entire cost improvement effort will receive a tremendous boost. Above all, management should take a constructive, not a defensive attitude. This may seem rather nebulous as a suggestion, but in the experience of the author it is absolutely essential to success.

2

Management Participation and Support

It is a truism that no one is "against" cost improvement. Unfortunately, when the real pressure is applied, it often turns out that everyone favors cost improvement in the abstract and for the other fellow. Cost improvement, as has been pointed out, represents change, and it also is seen as an unknown, as something that threatens whatever comfortable modes of operation have grown up in the absence of standards and goals. This is particularly true in indirect-cost areas. Line supervisors and middle management know that the only way they can get into real trouble with the boss is to fail to meet their functional responsibilities. They are much less likely to get in trouble because they are overmanned, since no one else knows how many people they really need. This may be overstated, but not by much. The same situation may hold in the direct-cost areas, if standards have been allowed to loosen over the years. The point of this discussion is that top management should not count on the thrust for cost improvement coming from the lower levels of the organizational structure. At these lower levels, cost improvement brings with it new problems. The thrust for improvement must come from the top. Cost improvement programs fail not because of a lack of techniques and procedures, but because of a lack of strong, visible top management support. In this chapter we shall discuss some specific actions and attitudes that have been successful in making this support both evident and effective.

THE DIRECTOR OF COST IMPROVEMENT

Cost improvement programs usually are organized so that line management is directly responsible for meeting goals, staff management for providing technical assistance, and a director of cost improvement for the administrative details as top management's representative. This director of cost improvement is a new position, but it should be filled by a present employee. He is formally responsible for the administrative details and informally responsible for being helpful to both line and staff in meeting departmental objectives. To be explicit, his formal responsibilities include:

1 / Communicating the statements of goals.
2 / Supervising the issuing of progress reports.
3 / Arranging the various meetings incident to the program.
4 / Helping where he can to coordinate line and staff activity.
5 / Publicizing the program so that all employees know the nature and intent of what is being done.
6 / Serving as executive secretary of the cost improvement committee.

Setting and Communicating Goals

There should be a very clear and specific statement of the goals for each department. These goals are usually set by a committee, and must then be issued in a standard form. At this point, the reality of the cost improvement program may set in for the first time, and supervisors deserve a chance to discuss the goals with a knowledgeable person. Despite the professed distaste for "paperwork," it is necessary that some form of statement of goals be maintained, consistent for all departments. Written ground rules for establishment of goals and evaluation of improvements are required, and the director of cost improvement should be responsible for these. Reducing these matters to writing and discussing them with each department head will also bring out whatever initial conflicts may exist; these should be brought to the cost improvement committee for resolution.

Progress Reports

As a corollary to the statement of goals, a consistent rationale should be developed for the validation and reporting of improvements. Here the need for specific ground rules becomes critical. There is a natural tendency to overstate gains, particularly if it is felt that this will reflect favorably on the department. But the integrity of the program must be maintained. We cannot afford to have anyone deprecate the results of the program on the grounds that improvement is overstated, or is measured in "Chinese dollars." Measuring results will be discussed later, but the basic policy should be to use the ordinary accounting practices, and to allow as firm savings only those improvements reflected in the regular operating reports that management now uses and is accustomed to. Where standard unit costs are available, this is a fairly straightforward matter. Where units of output have not been defined, however, evaluation of improvement is more difficult. The effect of this difficulty is to give impetus to the requirement for input-output measures that extend to all operations. Part of the benefit of the formal program is the establishment of these measures, particularly in indirect-cost areas such as maintenance, clerical, and service activities.

Meeting For Review

The statement has been made that top management must have a visible and continuing interest in the cost improvement program. One way to demonstrate this interest is with formal cost improvement review meetings. These occur monthly, quarterly, and annually. The director of cost improvement is responsible for making the arrangements and for chairing the meetings. This is almost routine. But the informal part of his job, which is more demanding, consists of going over in some detail each supervisor's presentation of results. This should be done in advance of the formal meeting to give the director an opportunity to be helpful on an off-the-record basis. Above all, we do not want the supervisor to fail or to be made to appear inadequate. We are less interested in reasonable excuses than we are in posting favorable results. Particularly at the start, we should remember that we are dealing in change, and we should make it easy for individual supervisors and managers to contribute to this change. The meetings themselves should be well organized, have an agenda, and stay within reasonable time limits. The quarterly and annual meetings should be viewed as significant events of real interest to all levels of management.

Coordinating Activities

The informal responsibilities mentioned in the last paragraph are critical to the success of the program. Many managers and supervisors will be suspicious of the program. If top management has done a good job of indicating strong support, the supervisors will be anxious to succeed. But the management process requires that we translate objectives into short-term goals and generalities of cooperation into the specific development of personal relationships among line and staff departments and among line departments. There will be many questions concerning accounting systems, analysis techniques, the job security of employees, and the appraisal of individual performance. The director of cost improvement can perform a most valuable service by talking with managers and supervisors, making phone calls, getting people together, and obtaining resolution of questions when necessary. This activity sounds rather redundant, because supervisors are supposed to do these things themselves. But we are anxious that the program work, and we want to remove every possible source of delay and misunderstanding. Only the naive expect changes of this magnitude to be made easily and without personal problems; the informal part of the director of cost improvement's job is to anticipate these problems, and prevent them from becoming critical.

Publicizing the Program

The director of cost improvement is also responsible for publicizing the program within the company. It is important that all employees know that their managers and supervisors are heavily involved in the program, and that there is real pressure for improvement to which everyone must respond. The most favorable honest statement should be made concerning job security. The objectives of the program must be communicated in a positive manner that stresses the pervasive nature of the effort and the existence of strong management support. There are many useful avenues of communication within a company, and the director of cost improvement should work with the personnel office and line management to select those most suitable. We must not forget that one avenue of communication is the "grapevine," and although this is efficient, it is not the primary source to which we should turn for information. Management must present cost improvement in a constructive way, and it should seek the widest exposure in communicating the message.

Cost Improvement Committee

The last responsibility for the director of cost improvement is to serve as executive secretary of the cost improvement committee. This committee is small and composed of the principal senior managers. It is not the director of cost improvement's responsibility to direct the work of these people; they are probably senior to him, and the cost improvement program as it applies to their departments is definitely their responsibility. The director of cost improvement conducts the meetings and makes sure that the administrative details are taken care of. He is in somewhat the same position as the director of safety, in that the various department heads are responsible for safety within their departments but the director of safety provides technical expertise and administers the program. It is obvious that the informal aspect of administering the program places real demands on the capacity of the director of cost improvement.

Selecting the Director of Cost Improvement

The preceding paragraph gives an indication of the delicate feel for personal accommodation that goes with the position of director of cost improvement. It will be perceived immediately, particularly by those with business experience, that this is a key position and vitally important to the cost improvement program. What type of person should we look for to fill this job? What should be his background? And what should be his status in the organization?

The first requirement for the position of director of cost improvement is that the person we are looking for already be employed in the organization. Since so many of his specific duties involve communicating with others, formally and informally, he should know many of the people with whom he will work. Next, he should be generally regarded as having a promising future in the organization. Educational requirements are nebulous and are not as important as an earned reputation as someone who gets things done. Directors of cost improvement are selected from such diverse positions as assistant controller, assistant plant manager, chief industrial engineer, and manager of manufacturing services. This does not exhaust the list; but no matter who is selected, it usually is true that he will be difficult to replace in his present job. At the same time, however, he will probably be the type of person for whom a move to new responsibilities will be acknowledged to be almost inevitable. Top management will face its first test of commitment in the appointment of the director of cost improvement. If an outstanding person is appointed to this job, everyone will realize that top management is serious about the program.

The director of cost improvement should be in a staff position reporting to either the chief operating officer at a location or to the second in line of authority. He should report no lower. He will need secretarial help, and that is about all. He should not have a staff himself, and should spend most of his time talking to people, analyzing reports of cost improvements, and setting up line-staff groups for departmental and interdepartmental projects. (Much of the work is done in such groups, in project form, so he should develop project-reporting documents.) He does not speak directly for top management, and should be careful to make clear to all department heads and supervisors that it is *their* program. In other words, he should emphasize that the game is for the players, not the scorekeepers.

The director of cost improvement holds a tough job. We hope to select an outstanding man initially, and to give him the opportunity to get very wide exposure within the company. We will be taking him from a job in which he has experienced success, and putting him in a new job for which some responsibilities will be defined only as the program takes form. But if management selects a first-class person for the job, and if that person lives up to expectations, the position of director of cost improvement should become most sought after as an opportunity to assume higher management opportunities. Conversely, the position is a challenging one, and should be regarded by top management as one in which an appraisal can be made of an individual's potential for advancement. But, again, the primary thrust should be toward selecting a person who can help make the program effective now. This takes a high-quality person of a type scarce in any organization. Top management can demonstrate its commitment by selecting such a person as director of cost improvement.

TOP MANAGEMENT SUPPORT

The executive's support of cost improvement will be judged more by his actions than by his words. The objective of the program will of necessity be in words. The program itself consists of a series of specific tasks that are undertaken by each supervisor or manager. Examples of these tasks are budget analyses and definitions of work units of output. In other words, the program is not presented as a flat "improve costs" directive, leaving the supervisor to wonder what he is supposed to be doing, specifically. Instead, the process is done step by step, following the conventional management pattern, and progress is appraised at each step. For example, in one step the supervisor is asked to make a Pareto analysis of his expense budget (this means arranging the items in descending order of importance). This is then gone over, discussed with him, and questions resolved. He then is ready for the next step. The supervisor is thus presented with a series of short-term assignments that are within his capacity to accomplish. In brief, it is reasonable. Management must then insist that these steps be done, however. Taking several well-defined steps that are capable of audit serves to identify difficulties as they arise. Furthermore, the supervisor should feel that he is in a game that he can win, which is encouraging. We are all familiar with the concept that we can overcome large problems by dividing them into smaller, more manageable parts.

It is quite possible that this insistence on cost improvement performance will be tested early in the program. For example, an experienced supervisor who is also under pressure to meet difficult production goals may protest to management that he has always understood his job to be "getting out the work," and not "fooling around with cost improvement." If management tells him to "do the best you can," he will opt for the meeting of schedules and ignore cost improvement. Management should insist on both production and improvement, although full assistance should be offered. The management position should be that cost improvement is part of the supervisor's job and to be expected as a matter of first-class performance.

Perhaps the most direct way in which top management can show its dedication to the cost improvement program is to make performance in this program a factor in the evaluation of overall supervisory performance, and relate it to salary administration. It should not of course be the only factor, and the exact manner in which this is done should be developed carefully to fit into the existing compensation plan. For although it is true that job security and self-fulfillment are major factors in job satisfaction, it also is true that money is extremely important; in many cases it is the overriding consideration. The cost improvement program does not envision immediate cash awards for each improvement, as is the case with suggestion plans. But performance in cost improvement should be recognized at the periodic performance reviews for merit raises and promotion. Again, if top management feels strongly enough about the importance of cost improvement, it should make this known through the direct and compelling message of salary administration. However, it is of vital importance that the program be well run and regarded by the supervisors as eminently fair and reasonable if it is to be used in appraisal.

Finally, top management must devote some of their time to the program. This will involve attendance at cost improvement program reviews and cost improvement committee meetings, and time spent in reviewing formal reports and in resolving the conflicts that arise. This may seem a burden, but at least the questions will arise within a framework of systematic progress toward a worthwhile objective. A collateral benefit of the program is that fundamental questions arise naturally, and that in answering them top management has an opportunity to handle them in a businesslike fashion, and not simply as a reaction to a crisis. In too many cases cost improvement is stressed only when unfavorable business conditions and a bad operating statement make cost really critical. Such crises come and go, but commitment by management to a continuing program will put in perspective for everyone the advantages of managing positive effort rather than reacting to negative influences.

In summary, the active support of top management is the one essential element in successful cost improvement programs. In this chapter we have covered some of the details of form by which this support is reflected. These details are not characteristic of all programs, of course, but the need for strong conviction and articulate statement by top management is imperative to success.

3

Considerations of Behavioral Sciences

HUMAN FACTORS IN COST IMPROVEMENT

The previous section has discussed the need for the involvement and support of top management. The point was made that one of the basic reasons for cost improvement programs is that such programs enable top management to apply the concept of a formal structure which includes literally everyone in the operation. The point is made also that such programs allow the manager to exert his skills in the area of cost improvement. The implication here is that the manager will achieve results in this area through his people. This is, after all, one of the definitions of management; that is, "management is the art of achieving desired objectives through the coordinated efforts of people." So there should be no inference drawn that cost reduction programs are rigid and authoritarian in the way in which they operate. As a matter of fact, we have termed such an authoritarian approach as the "meat-axe" approach, or "management by fiat." Furthermore, this authoritarian approach was presented as an alternative to the cost improvement program which is suggested for use. However, there is a fundamental aspect of any program undertaken as a management activity which deals with the question "What are the behavioral science aspects of the program?"

It is appropriate to introduce this particular question as a corollary to discussion of top management responsibility. The traditional approach to cost improvement, it must be confessed, has been authoritarian in nature. This has been true for a number of reasons. First of all, the concept of using the budget alone in an arbitrary fashion to control costs is acknowledged to have been successful in the short term in many instances. Secondly, it has not been until recently that the manner in which management seeks its objectives is of great importance. Even the compelling evidence of the Hawthorne studies, which were conducted in the 1930s, seems to have been largely ignored in the practice of management during the following two or three decades. But if management in general was not perceptive about the Hawthorne studies, professionals in the fields of psychology and other disciplines dealing with human behavior saw in these experiments proof that there was opportunity to extend to the practice of management concepts which had been useful in dealing with interpersonal relationships. For this we should be grateful. The increased acceptance of the concept that individuals respond to a recognition of their worth as human beings and that financial reward alone does not necessarily insure high motivation was not long in coming. It is easy enough to say that these concepts were not new, and to point to the work done at the turn of the century by people such as the Gilbreths, but the fact is that for many managers it came as a shock to find that approaches other than authoritarian, money-oriented motivations were, in fact, most effective in some cases.

Current Theory Applied to Cost Improvement

At this point the assumption will be made that the reader is familiar in a general way with some of the thinking and writing in the field of behavioral science. Such concepts as Frederick W. Herzberg's classification of factors affecting the motivation to work as "dissatisfiers or hygiene factors" and "satisfiers or motivators" and Abraham Maslow's definition of the hierarchy of human needs are now common knowledge. It does not seem appropriate here to enter into any extended discussion of the literature of behavioral science. There is, however, one concept which seems to epitomize the difference between the arbitrary "meat-axe" approach of traditional cost improvement and the organized program suggested in previous chapters. That is, Douglas McGregor's approach in which he characterized the practice of management as following either Theory X or Theory Y.

McGregor characterized the assumptions underlying much traditional management practice, as he had observed it, as Theory X. He defined these assumptions, which were essentially negative in nature, as reflecting a belief that the average human being not only has an inherent dislike of work but in fact looks for direction from his superiors and values security above all. Theory Y, on the other hand, which McGregor felt to reflect the result of research in the behavioral sciences, was, as stated, a group of concepts which advanced a more modern approach. Briefly stated, Theory Y takes a more positive view of the worker—it states that we can assume that the average human being does not inherently dislike work; that work may be a source of satisfaction; that creativity in the solution of problems is widely, not narrowly, distributed in the population; that rewards in the area of personal satisfactions are important; and that the extent to which men and women are committed to personal objectives really governs their motivation to work. These concepts are discussed extensively in McGregor's book, *The Human Side of Enterprise,* published by McGraw-Hill in 1967.

If we accept McGregor's work as significant and representative of good professional thinking in the area of behavioral science, there are two points which should be made. First, there is an obvious application of this theory in examining the differences between the "meat-axe" approach of rigid, budget-controlled cost improvement programs and McGregor's Theory X. The flat budget cut, for instance, is seen by most people to be an indiscriminate strategy which presupposes that the people to whom this is applied are inherently unimaginative and are working below their capacities. This assumption, of course, is justification to the arbitrary nature of the reduction demanded. This seems to be pure Theory X. In this approach, the manager is assumed to have been working well below his capacity, and the directive to him reflects this assumption.

On the other hand, the assumptions of Theory Y seem to be more consistent with the pattern of cost improvement given in this section of the book. In this approach, the assumption is that managers at every level have both the initiative and the ingenuity to improve, and the program offers them the opportunity to set goals, receive credit for their work, and influence the behavior of others working as a team toward the realization of these goals. In fact, it may be said that a systematic cost improvement program provides a framework within which it becomes socially acceptable for an individual to exercise his ingenuity and motivation. In return for this, he receives recognition and some financial reward. (It should be noted that financial reward alone is not as great a motivator as financial reward plus satisfaction for meeting personal goals.) These characteristics seem to be consistent with elements of Theory Y.

The second point which must be made is that human beings differ widely, and while it is perfectly true that the Theory Y approach seems to appeal to those people whose psychological make-up responds to a less authoritarian approach, it is also true that there are many people whose behavior patterns seem to make the assumptions in Theory X thoroughly valid. In other words, people who respond to Theory Y may be a minority. It still seems to make more sense to appeal to these "Theory Y" people, however, because in the long run they are the ones who will be most valuable in instituting change and responding to challenge.

Recognizing Existing Management Attitudes

It is an obvious statement, therefore, to say that the behavioral science aspect of introducing change is extremely important to the success of any cost improvement program. What is not so obvious is the course of action which an organization should take to insure that principles of human behavior are taken into account. One thing seems obvious; that is, that any approach which seems forced, artificial, or not consistent with general management tactics now existing in the organization will probably be recognized as being "phony" and will do more harm than good. In other words, if the prevailing management style is pure Theory X, introduction of a cost improvement program which is based strongly on many of the assumptions of Theory Y will be perceived correctly as somewhat of a contradiction in management style. It is true that a cost improvement program offers a splendid vehicle for transmitting management's attitude, changed or not, and indeed such a program gives management a great way to introduce a genuine realization of the importance of human behavior. But above all, this must represent a sincere conviction and the willingness to change. If the people whose support is being sought come to regard behavioral science as a gimmick which management is using to manipulate them, not only the program but management's credibility will be destroyed. It simply is not possible to patch over serious deficiencies in the area of human relations with any program. This is particularly true of cost improvement programs, since their effect involves everyone in the organization.

Keeping in mind the caution of the last paragraph, the author does not suggest any particular "human relations" formula. There will be much discussion of the specifics of managing such a program and of involving people at all levels who will in effect manage themselves. But there will be little discussion of personal style in human relations. This is because whatever approach is taken must be genuine and natural to those within the organization. Every company has a style of management. Furthermore, there is a unique history of human relations within each company. Some companies already have recognized the importance of the behavioral sciences, and have been working rapidly to improve the interpersonal relationships existing within the organization. In such cases, introduction of a cost improvement program of the nature discussed in this book will be relatively easy. Other organizations, which are accustomed to receiving all direction from the top, will find it more difficult to introduce a participative cost improvement program such as this book proposes. This will be true not only because this program represents a departure from existing authoritarian practice, but also because the people at the lower levels in the organization will have to be convinced that creativity and initiative are in fact organizational goals and that they themselves are expected to exercise these qualities. This may represent an almost overwhelming problem in changing attitudes. But the optimistic view still should prevail. If organizations can be brought to consider seriously the implications of behavioral science in the cost improvement program there will be a better realization that the same principles apply in all organizational matters. This can only be constructive.

Finally, the reader should understand that this discussion of behavioral sciences as it relates to cost improvement programs is by no means a definitive one. However, there is an extensive body of knowledge which has emerged over the past few years, and the author is confident that the reader has some familiarity with the work of people such as Gellerman, Herzberg, Lehrer, Likert, McGregor, and Roethlesberger. While their ideas have been advanced over the past few years, the problems have been with us for a long time. Perhaps the best single comment which we all should keep in mind is the perceptive observation of the Swiss playwright, Max Hirsch, who said, "We sought workers, and human beings came instead." Enough said.

4

Effects and Problems of Companywide Application

One responsibility of the top management executive is to be the final arbiter of differences of opinion that arise anywhere within the organization. In settling these, he not only evaluates the different points of views, but almost always one department's view is upheld and another at least partly rejected. The top manager must consider the entire operation, however. He realizes that such decisions resemble golf in that "every golf shot makes someone happy." The corollary is that someone else is unhappy. The executive's problem is to make everyone see that there are overriding corporate goals, even if the day-to-day work may seem to center on more narrow departmental goals. In introducing a cost improvement program, the top executive is in effect forcing each department head to work toward both departmental and company goals; this is best made clear if everyone knows that all departments will be included, eventually, and that every department will be asked to perform in the same way. Although it may be necessary to cover the entire organization in progressive steps, it is important to have companywide scope. The basic reasons for this are that (1) it is the fair way, (2) many significant cost improvements require cooperation between departments and the program encourages this, and (3) everything can be improved, so we should make no *a priori* judgments that any department can be excluded.

The first reason is most important. If, for example, you were in the production department and had been included in the program, but the controller's department had not, what would your reaction be? The answer is obvious. Management is going to insist on results and will expect a positive attitude down the line. This does not leave room for the question "why me?" Management will save itself a great deal of frustration if it makes the point that everyone is expected to contribute. This is obvious, and needs no further discussion.

INTERDEPARTMENTAL COOPERATION

The second reason involves us in the mechanics of the cost improvement program. Each department has a goal to meet, but some improvements affect more than one department. If there is no cost improvement program, there is no well-defined way to recognize the contribution of each department when a joint effort results in cost improvement. In this case, a department head may be reluctant to assume responsibility for a new procedure that should result in an improvement which will show up in another department's operating statement. In brief, he will ask himself why he should accept a problem when someone else will get the credit. If there is a cost improvement program in effect, however, the credit will be shared. This is perceived as fair by the supervisors, and opens up the entire operation by removing a very real inhibition to working across departmental lines. Anyone who has had experience in management at any level knows that self-interest is a fundamental human characteristic; indeed, we appeal to this characteristic in wage and salary administrative practices, in merit rating, and in the establishment of a management hierarchy for upward mobility. Rather than expressing a pious hope that people will put aside personal ambition in favor of company objectives, we should make it possible to achieve both. This we can do through a cost improvement program.

As a specific example, in a large printery the practice for years had been to buy five-foot-square sheets of light cardboard for the covers of clothbound books. These were stored in the preparation department. When the bindery (which actually used the cardboard) needed the material, the supervisor would call up the preparation department and simply ask for the right amount of the proper (book) dimensions. He would also give a delivery time, usually the next day. (Sometimes the same day, of course.) The preparation department supervisor would have the large sheets cut to book size on a huge paper knife, piled on pallets, and sent to the bindery. One operator per shift was employed to do the cutting, although his utilization on this job (and the utilization of the expensive equipment) was quite low. Furthermore, the scrap rate from cutting was high. For years, the purchasing department had known that the cardboard could be purchased from the mill already cut to book size at a slight increase in cost. Delivery time for the mill was one week, and the demand could be forecast quite accurately about three weeks before the covers would be needed. However, the bindery supervisor liked the convenience of just calling up for stock, and the preparation department supervisor did not object to the arrangement, since he could use the operator who did the cutting as a general errand boy (although a high-priced one) in the operator's considerable spare time.

This sounds like a loose arrangement, as in fact it was. But the reader should not dismiss this as an atypical "horror case." These things happen all the time. With no specific pressure for cost improvement, such comfortable arrangements grow up almost naturally. In any event, a cost improvement program was started, and the resulting systematic analysis of budget and work units brought out the situation in the bindery. At the same time, all three departments involved (purchasing, bindery, and preparation) were feeling real pressure for improvement. So the matter became a project, the solution to which was to purchase the cardboard already cut, to have the bindery take the responsibility for ordering three weeks in advance, and to eliminate three jobs in the preparation department. The three employees were reassigned, at no loss in pay, to fill open personnel requisitions. (This last was possible because the company had gotten a large new contract, which made the change much easier.) The net savings, amounting to about two payrolls, were split among the three departments.

The essential facts here are that the improvement could have been made years before, that the cost improvement program did not contribute much in the way of technical assistance, but that the program did provide the stimulus to improve. It furnished a means of "horse trading" so that all three departments had a vested interest in seeing to it that the change was made smoothly and that it would continue to work. It is obvious that this should happen without a program, but the fact is that it simply does not happen without some pressure on the part of management.

Many supervisors are reluctant to take actions that cut across departmental lines, because the traditional pattern has been to stress these lines as a means of maintaining orderly management. It is difficult to relinquish one's department-oriented self-interest for the "best interest" of the organization as a whole, because the definition of these organizational "best interests" may depend upon who is doing the defining. If we can recognize joint action by giving credit in a manner consistent for all, the "Chinese Wall" syndrome will be less evident. Cost improvement programs use the device of sharing credit to make it easier for departments to work together.

JOB SECURITY

One aspect of cost improvement that must be dealt with directly is the question of job security. This is in a sense a negative rather than a positive factor. But each individual in the organization has a very understandable interest in his own job, as it is now, and in his prospects for the future. Cost improvement programs are disquieting in that the fundamental need for and benefits of such programs to the company as a whole are harder for the individual to visualize than is the fact that such programs actively seek change. Employees can visualize change as it applies to their jobs, and to many it represents a threat to their job security. This is certainly true in the sense that habit patterns of many years' standing will probably be altered. Particularly in indirect areas of operation, analysis for cost improvement often reveals situations that are clearly overmanned. The resolution to this type of problem is almost bound to involve changes in job assignments at the least; if no new work is found to absorb excess manpower, these could be severe.

In Chapter 2 the statement was made that management should put together the most favorable honest statement of job security which it feels can be made. Examples of such statements might include the following:

1 / No one will be required to move from his department; any improvement will be reflected in the assignment of additional workload, or in not hiring to fill vacancies. Retraining of personnel will probably be necessary.

2 / No one will lose his job, but there may be reorganizations that will cause personnel transfers at no loss of pay to the individual. Retraining will probably be necessary, and the improvement may also be reflected by the assigning of additional workload and in not hiring to fill vacancies.

It is obvious that these two typical statements reflect different degrees of job security. When possible, management would prefer to give a firm commitment to a statement such as the first, because it is the most favorable. But the fact is that unless there is increased output in some form, offsetting gains in material usage or substitution, or the development of new products or services, cost improvement may require the reduction of personnel costs. Job security of course varies with factors of size, skill of the individual, and market conditions. For example, there is no lack of security for a skilled maintenance mechanic who now works with overtime in a process industry that is booked to capacity for the next year. (We might try to reduce the overtime, however.) There might be less job security for an equally skilled typist working in the housing industry, which is currently in trouble because of greatly increasing material costs and difficulties in securing home financing. Whatever the situation, management should realize that a drive toward cost improvement should be accompanied by persuasive publicity about the fact that real job security depends first upon the economic success of the company; without the ability to compete in the marketplace no company can offer any job security at all.

The point should be made that options exist for the individual affected by change. Thought should be given to retraining and to reassignment within the company. While the employee can reasonably expect that the company will make every effort to keep him on the payroll, he cannot reasonably expect that he will be retrained on the same job, done in the same way, when change occurs. Technological improvement is a fact of life, and everyone's job changes, including the manager's. So we can require shifts in job content and departmental assignments, provided we are fair about it.

For example, in one pharmaceutical manufacturing plant, a cost improvement program resulted in the elimination of nine skilled jobs in production. At the same time, the finishing (packaging) department employed about forty part-time operators on a second shift. These operators, usually off-duty police and firemen, had been hired with the understanding that the work might not be steady, and that the same degree of job security enjoyed by regular employees would not be extended to their jobs. The nine skilled operators were transferred to the finishing department, with no loss in pay and with the assurance that they would be returned to more skilled work as jobs opened up in the future. Some objected to this, but were told that the alternative was to be laid off. The company's position was that the new arrangement enabled the employees to retain their pay rate and their seniority, and that they were in fact being paid more than others on the same job. This was accepted, and no problems arose.

In another case, tabulating machine operators were retrained as computer programmers. This cost the company money, but it made the cost improvement program much more acceptable to all the employees; they felt that this was proof that management was committed to improvement, but not at the expense of reasonable job security for their employees.

One other aspect of job security is the fundamental economic truth that job security is associated with a growing, successful company. There are really two ways to absorb the excess personnel and plant capacity that may be created by the working of a cost improvement program. One is natural attrition, which is effective but not necessarily reliable. The other is to devote the resources freed by the program to produce more product, to improve the quality of product or service in places where there is prospect of return, or to develop new products or services. A common practice is to take some part of cost improvement gains and devote them to special sales promotions. Here, the concept of incremental cost becomes important. The sales force is always on the lookout for special promotions and special services. Unfortunately, these may come at the expense of operating economy. But if the people and facilities become available through more effective operation, management may feel more inclined to take the risk. The same is true in manufacturing.

In one case, labor made available by cost improvement was used to salvage rejected assemblies. This had not been thought to be worth the production interruption, which would have been necessary had the salvage been done as scheduled work, but it turned out that the salvaged material more than repaid the cost of labor. The key here was management's taking a positive approach to job security by trying to expand opportunity, rather than the negative approach of giving job security to those who are left as attrition reduces the work force.

COMPANYWIDE APPLICATION

When the cost improvement program is introduced, it should be made clear that all parts of the organization will be included. But, at the same time, the staff assistance required to cover the whole organization may not be available. Or management may feel that it is more effective to concentrate on a pilot-study procedure. Although it makes sense to start everyone at one time, this is not always feasible. It also should be kept in mind that when a program starts there are unusual demands on the time of staff and supervisors because the program is new, if for no other reason. A real choice may exist as to whether to start all or only part of the departments on the program. If only part are started, there must be a firm schedule for the rest. Usually the time differences involved are of two or three months' duration. This is true because the initial activity may involve a month's work sampling, which will involve analysis of a month's data on production or work units of output.

If the decision is to start with only part of the organization, thought should be given concerning which part. Usually we start with at least two departments and no more than four. (The last restriction presumes lack of staff.) The definition of a department varies widely, but we suggest that the scope of a "department" cover from 12 to 100 people and from one to perhaps six or eight supervisors. This is, however, a matter of local judgment. There should be enough variety to form a representative sample, and enough organizational coherence so that channels of responsibility and accountability are relatively straightforward. The departments should already have budgets and operating statements so organized that we can identify both expenditures of resources and units of production or output by departments. When there is any choice, departments with substantial expenditures should be selected. Selection will be made on the basis of the first rough budget analysis. This criterion is sensible, since management should want to expend its effort where there is most opportunity for gain.

A final criterion is to select departments to start the program that include the strongest supervisors and which seem to be "running well" now. There is a temptation to use cost improvement programs to "clean up" unsatisfactory departments. These attempts usually fail. The positive reasons for starting with strong supervisors and satisfactory departments are the following:

1 / To be practical, the strong supervisor has already shown that he can handle change. So why not use this demonstrated capacity rather than risk the program by putting it in the hands of a supervisor who cannot even make the present system work?

2 / If by some miracle we did get some improvement, it would be deprecated as meaningless because "the operation was so bad that *anything* would be an improvement."

3 / The strong supervisor will be better able to make the program work,

of course, but will in addition see in it an opportunity for self-advancement. This we encourage.

4 / When savings are made in departments acknowledged to be well run now, this reinforces the notion that almost everything is subject to improvement.

5 / Finally, the strong supervisor is more likely to bring to management's attention the difficulties that are bound to arise. This feedback is necessary, and the good man will be more likely to regard the difficulties as simply problems that can and will be solved. The weaker supervisor tends to regard the same problems as insurmountable roadblocks that should cause immediate abandonment of the program.

To summarize, the coverage should include everyone, eventually. If the program is started in but a few departments, everyone else should know that he is scheduled for inclusion. Strong departments with good supervision should be chosen to start, and significant resources should be represented to avoid wasting effort.

The advantages of companywide scope are that it is seen by the employees to be fair to all, it allows different departments to share a single project, and it helps remove the reason for some of the human self-interest which inhibits any change. Finally, there is the self-evident advantage of having all managers working together under a single set of rules to achieve a worthwhile companywide goal.

5

Line Responsibility

The process of management requires that channels or lines of authority and responsibility be assigned, and that objectives be transmitted and results evaluated following these channels. In the typical business or institutional organization, we designate those positions in the channel of direct responsibility to accomplish the various main functions of the business as *line*. Positions in functions that are not central to the main thrust of the business are designated as *staff*. The reader is undoubtedly familiar with this concept. An organization chart of a typical manufacturing company shows these relationships clearly. As a generality, positions such as president, vice-president for sales, vice-president for manufacturing, general manager, division sales manager, general foreman, and foreman (first-line supervisor) are line positions. Positions such as vice-president and chief counsel, vice-president for management services, chief industrial engineer, personnel manager, system analyst, or safety engineer are staff positions. This is all subject to specific organizational goals, of course, but for a specific organization the positions are easy to classify.

In cost improvement programs, it is essential that responsibility for improving costs and making the program work be assigned to the line structure, except as internal goals are assigned to staff functions. The usual focus of attention in cost reduction efforts is the production floor. It is absolutely true that here the first-line supervisor or foreman is the key man. The concept of

line responsibility means that on the shop floor responsibility for cost improvement lies with the foreman rather than with the industrial engineer. They must
work together, of course, but the first-line supervisor is accountable. But we
should not really start with this level of the organization. The president is in the
line, and he bears the ultimate responsibility. So although it may be traditional
to start with the foreman, we should start at the top.

DUTIES OF THE PRESIDENT

We have described the president's obligation to give support to the cost
improvement program, but he should not limit his participation to these duties.
He himself should take the position of looking at all costs, external as well as
internal. At times he should take the point of view of the customer, on whom
the success of the company depends. It has been pointed out previously that the
customer regards a company as just one of many possible vendors, no matter
what the product or service. This should be stressed over and over. If the president limits his concept of cost improvement only to those things which he controls, he is not providing leadership. In Chapter 4 the point was made that every
area should be covered, and that some functions probably should be changed
radically or even eliminated. It was also stated that the cost improvement program provided a means of appraising cost versus benefit (input-output) so that
problems could be approached within a more objective framework. The president is the only one who can demand an open spirit of inquiry from everyone.
One of the hard tasks of leadership is to balance the reliance on the familiar,
which is necessary for day-to-day operations, with a positive pressure for change.
The traditional areas of cost improvement still have potential, of course, but the
percentage of cost represented by internal expenditures has been declining. Now
transportation, financing, purchasing, distribution, and government-imposed
costs are becoming more critical, and these are largely influenced by forces
external to the company.

This last point is made not to minimize internal cost improvement, but
to emphasize that the president is in the top line position, and he has the unique
responsibility of directing the total effort. He is the one who must take the
external view, as well as the internal. Most of our discussion will be in terms of
input (cost) versus output (product, service, or work unit) analysis. The ultimate
question is whether or not the output is attractive to a customer or is worthy of
public support. It is the chief operating officer's job to represent the customer's
viewpoint.

Many companies use the term "profit improvement" or "profit plan" to
describe programs involving cost improvement. This does give a broader implication, because it is fundamental that cost improvement has to increase profit.
But there are so many factors in determining profit which are beyond the control of the department manager that he sometimes does not see it as realistic

to use the term. Many of these factors, for example, are external, and only the president is responsible for all of them.

What, specifically, should the president do in achieving cost improvement? First, he should not destroy his organization by trying to interfere with the responsibilities of his senior managers. He should read the appropriate periodicals that report on his industry, on industry as a whole, on governmental operations, and on the financial world. He should talk to his senior people and keep in touch with what his competitors are doing. He should talk with his customers, his banker, and his lawyer. Much of this will be nonproductive, of course. But here and there he will pick up the germ of an idea that will make it all worthwhile. Then he should "needle" his own people with these issues. Even though there may be some lost motion and even resentment, it is well to establish this as a practice. Finally, questions of basic policy that involve cost improvement are brought to him; this gives an opportunity for broad discussion and questioning. Adherence to orderly procedures for administration is important, but there are advantages to an open and freewheeling approach to special problems. The president can use the cost improvement program to stimulate such an approach, because he may define operational areas about which he has some misgivings as special problems. But he should not be only an administrator of the program— the director of cost improvement has that job. He should be an executive, a leader, and a "mover and shaker" in the program.

DUTIES OF GENERAL MANAGERS

Managers at the next level in the line should also participate actively by going over the operating budgets for each department and, above all, by reviewing the outputs of each function for which they are responsible. This level of management has somewhat the same position as the president in that they have a more comprehensive feel for the customer, the vendor, the money market, or the technology. At this level of management there should be a broad appreciation of the enterprise as a whole. The basic decisions made at this level have to do with suggestions for the thrust of cost improvement and the allocation of resources of time and people. These executives should of course meet with their own people for a free and open discussion of priorities and allocations. But above all, they should take a real interest and become involved themselves. The cost improvement program should not be regarded as simply another chore to be performed at the lower levels of the organization. Instead, it should be taken as an opportunity to make changes that should eliminate marginal carryovers from the past and to build for the future.

As an example, a manager of manufacturing in a carpet mill had an analysis done on his yarn supply. He chose to look at this because it was a substantial item of cost and because he felt no one had reviewed it critically; it was taken for granted. He found that he was doing half of his business in just four colors.

Yet the size of his production runs in these colors seemed to be about the same as those in his less popular colors. This was true because a "standard" size of dye lot had been accepted years ago. He worked back through the supplier to increase the size of dye lots in the popular colors, and thus effected substantial inventory and manufacturing economies. At the same time, he was able to cut down on the size of run in the less popular colors. It was found that carrying costs ate up savings of setup (tie-in) costs. This sounds like something which should have been done years ago, as in fact it should have been. But the fixed dye-lot size had become almost an article of faith, and had been regarded in manufacturing as a very rigid constraint. The fresh look generated by the cost improvement program, together with higher-level interest, provided the thrust to accomplish this long overdue change.

DUTIES OF DEPARTMENT MANAGERS AND
GENERAL FOREMEN

The next group to be involved in the line are the department managers or general foremen. On an organization chart these people appear as second-line supervision. Their responsibilities are narrower in scope than the groups discussed previously, and thus their opportunities for cost improvement seem more limited. But they should be the ones to whom the first-line supervisors turn for help to provide solutions to problems unique to the department. And since first-line supervisors deal with the largest groups of employees, it is essential that they get the help that they need from the general foremen and department managers. Department managers are of course able to supervise intradepartmental changes, but they also have the status to deal with other department heads. This the first-line supervisor does not have.

For example, the manager of the maintenance department in one process-industry plant was given a goal in cost improvement; at the same time, he felt that he was faced with other problems which were more compelling. Overtime hours were running at about 20 percent; even with the overtime, he had problems with expensive call-in work. The planning system was ineffective; first-line supervision was spread too thin, and the more experienced craftsmen would not accept promotion to first-line supervision.

The manager of maintenance did not recognize that the cost improvement program, far from being another irrelevant demand on his time, offered the means of solving his problems. He knew that the overtime was a root cause of his problems. He had presented a plan to reduce overtime, and use the savings to hire a few extra craftsmen and promote a few craftsmen to be first-line supervisors. His attempts to make the changes had met with failure, however, because of lack of support from the personnel department and the general manager to whom he reported. It was felt by these people that the work was getting done under the present arrangement, and that to reduce overtime would lead to other problems, which they did not want to face. When a cost improvement program was started, the manager of maintenance reduced the problem to a comprehensive proposal with cost as the criterion. Furthermore, he was able to support his proposal with a work-sampling and work-order analysis.

Since the general manager not only was responsible for the overall goal, but had also given the manager of maintenance the goal for maintenance, he was almost forced to support the manager of maintenance. The changes were made, but it was understood that the manager of maintenance would have to make the changes work. This he did. After a year, the planning system was effective, overtime was down sharply, and regular craftsmen had been put on shifts to cover work previously done on a call-in basis. Furthermore, with little overtime premium pay, the supervisory jobs were more attractive, because, when there had been substantial overtime, an experienced craftsman in some cases made more money than a first-line supervisor. Now there was much less difficulty in filling the first-line supervisory positions. The problem had been turned into an opportunity.

The preceding sentence is an old and sometimes bitter slogan. But the essential fact here is that the cost improvement program provided the framework within which a change could be made gracefully. Sometimes people take strong positions from which they feel that retreat would be an admission of weakness. The cost improvement program is a visible change in such a situation, and provides a face-saving device that makes change possible. This sounds unimportant to those who have not had experience in management, but in fact it is vital in maintaining the morale and usefulness to the organization of experienced people. After all, we seek to encourage the human characteristic of decision making, and just because some decisions are wrong we should not discourage future initiative.

This brings into focus the notion that giving the line the responsibility for cost improvement reinforces the established management structure. For in this case the actual day-to-day improvement came through the efforts of the first-line supervisors, who made the planning work, who assigned less overtime (and listened to the complaints which this brought forth), and who were responsible for getting the work done. They reported to the manager of maintenance, who had overall responsibility and who alone was in a position to effect the changes in policy that he had instituted.

DUTIES OF FIRST-LINE SUPERVISORS AND FOREMEN

Finally, we come to the first-line supervisor or foreman. He is in fact the key individual in his area in that he must make the change work. He must deal face-to-face with members of the bargaining unit, resolve the daily crises, meet production goals, and keep schedules. In addition, he has policy restraints to his freedom of action and the responsibility to operate many programs given to him by people outside his department. His ability to obtain improvement is limited by all these; yet, because it is he who has the primary responsibility in his area, he must have the same degree of responsibility for meeting cost improvement goals. It is his department or his group, and it is unthinkable that anyone else should have the authority to direct operations in this group. He will need extensive staff assistance, of course, and will work in some sort of team effort. But no staff personnel can or should infringe upon his authority to deal with his people.

With the increasing complexity of new products, processes, and personnel practices, the line-staff relationship has developed new problems. For example, the foreman looks to manufacturing engineering for process specification, to the toolroom for tools, jigs, and fixtures, to the industrial engineer for work methods, and to the production and inventory control group for schedules of production and availability of materials. This is as it must be to take advantage of improvements in technology. Yet this situation carries with it the seeds of frustration and a very real problem in the assignment of accountability. The foreman has become used to working with staff, but he has also learned to use any ambiguities of his particular line-staff organization to his own advantage when policies and programs distasteful to him are introduced. Most foremen and first-line supervisors are genuinely committed, as they should be, to the philosophy that their most important function is to "get out the work." This they try to do. However, foremen sometimes feel that staff members do not take the trouble to try to understand them or their problems. So the foremen rationalize as necessary to "getting out the work" the various stratagems that they sometimes use to exercise real control in areas where they feel that the "line" demands of running the shop override the "staff" requirements of specialized knowledge. This is a complicated question, but the problem exists.

For example, who has not seen all sorts of "spare" in-process components and parts being carried as protection against stockouts? And how often is it possible even to trace their origin? How often do foremen resort to "blending" off-quality material to get whole lots through inspection? And once a foreman has learned how to make a reporting system reflect the answers that seem to keep management happy and the staff out of his area, does anyone doubt that he will take advantage of this knowledge? On the positive side, the good first-line supervisor or foreman will perform wonders in material substitution, emergency repairs, ad hoc arrangements with his work force, and instant re-design of product in order to meet his production obligations. The real problem is to take advantage of the foreman's drive, motivation, and shop knowledge without giving him the feeling that he is being harassed by "another staff program." The only way to do this is to make cost improvement a line responsibility, although with staff support. When we think of it, there is no reasonable way in which we can or should make the staff responsible for the work of people under the authority of line supervision at any level. If we interpose staff direction of operations, the line supervisor cannot function effectively. We want the wage-roll people to look to their own supervisor for direction, and that is the way it should be. Frederick W. Taylor's concept of "functional foremanship" foundered on the rock of the hard fact that each employee should have only one boss. The first-line supervisor should have adequate staff support, of course (see Chapter 5), but the cost improvement program should be a line program, where we are dealing with line functions and line employees.

This is no departure from existing standard practice in other respects. For example, the line supervisor and the safety engineer work together on the safety program. But it is the foreman, not the safety engineer, who enforces the rules concerning wearing eye protection. And although the industrial engineer should specify motion pattern, tooling, and workplace arrangement, the foreman is responsible to see that these are maintained in the day-to-day operation of the job. Similarly, the foreman should accept staff assistance, but still should be responsible for meeting the specific cost improvement goals for his own part of the organization.

In summary, the line organization, which has the responsibility for the fundamentals of operating an enterprise, should also have the responsibility for cost improvement within the scope of their duties. This is particularly true at the higher levels of management, because the scope of responsibility is greater and thus the opportunities for improvement are greater. Furthermore, this arrangement forces consideration of matters of policy that otherwise might not be reviewed. As the level of management structure becomes lower, so too does the opportunity for cost improvement. Unfortunately, many traditional programs are directed only to the largest groups of people, and not to the largest items of cost. Thus, while it is necessary that the first-line supervisor have direct responsibility in his own area, all levels of management should be actively involved in the cost improvement program. To make the program only a first-line operation is to miss the point.

6

Staff Support

The discussion of line responsibility in the previous section emphasized repeatedly that there should be a concomitant staff responsibility to work with line supervisors and managers to provide more specialized technical knowledge and assistance. There are two basic reasons for discussing the point here, and for relating it to the operation of cost improvement programs:

1 / The institution of a cost improvement program places a new requirement on the line supervisor; to meet this, he almost certainly will need the help of the staff. We had better recognize ahead of time these new requirements, and work out the arrangements to meet them.

2 / In all too many cases the existing relationships among line and staff personnel are simply not satisfactory from top· management's point of view. In many companies and institutions line and staff do not work together very effectively (more about this later), and a cost improvement program may offer a new framework for improving this vital joint relationship.

TECHNICAL ASSISTANCE REQUIREMENTS

To deal with these in order, there are several areas in which the institution of a cost improvement program may result in a demand for either new or expanded line-staff cooperation. Usually the demand is for expansion, because we know that a fairly reasonable relationship should now exist. But at least the following areas of cooperation should be discussed:

1 / Need for technical assistance in instituting changes in the product, process, or nature of the service provided. Staff functions concerned here may bear titles such as design, manufacturing engineering, quality control, chemistry, metallurgy, marketing, or customer service.

2 / Need for technical assistance in formulating projects, collection and analysis of data, system design, and mathematical modeling. Staff functions here may be known as industrial engineering, information systems, operations research, or methods and standards.

3 / Need for technical assistance in the changes to equipment and facilities that may be required to implement an improvem t. Staff functions involved here may be known as plant engineering, facilities management, or maintenance.

4 / Need for technical assistance in the area of personnel administration. Many of the improvements made will involve changes in the number or qualifications of people, will have an impact on union-management agreements, may require changes in job assignments, and will surely raise questions about the underlying but compelling issue of job security. These matters are administered by the personnel function. Usual titles for this function include industrial relations, employee relations, or personnel relations.

5 / Need for technical assistance in the area of accounting practice and financial controls. This is obvious from the nature of the effort—cost improvement. The staff functions involved here may be accounting, cost accounting, or the controller.

It would be foolish to imply that these needs do not now exist or that the various staff functions involved do not now have established procedures for dealing with change. Sometimes these could be vastly improved, and this will be discussed later. But the institution of a formal cost improvement program does mean that changes will be sought more actively in a deliberate effort to meet program goals. This will result in more interaction among line and staff personnel, more time for meetings, more technical work by the staff, and more conscious attention on the part of line personnel. This may seem to be a burden, and it is absolutely vital that this total effort be directed only to projects which have a very good prospect of a significant financial payoff. Of course, we hope that some of these projects will also reduce "fire-fighting," which is a drain on everyone's capacities; but we should recognize that cost improvement programs require the allocation of scarce resources, and should plan for this. Our hope is that management support and a general acceptance of the compelling need for improvement will lead to the development of a much more efficient mode of line-staff operation. Experience has shown this to be a reasonable expectation.

Changing the Product, Process, or Service

The organizational relationship between line and staff can be stated fairly directly, and the distinction between "action and responsibility" and "advisory" is quite clear-cut in the abstract. However, reducing this distinction to the operational level can become a source of contention. For example, production methods designed by industrial engineering may not meet expectations as they are applied in the shop. Whose is the responsibility? Or a quality inspection deemed effective and valuable by staff quality control engineers may not be so regarded by the line personnel, and thus may be observed in the breach. Again, where lies the responsibility? Or a process that is effective under the controlled conditions of a semiworks or pilot-plant operation may not be effective at full-scale operation. Is the failure due to inadequacies in process design, or simply because operators are careless or the production equipment is incapable of holding process limits? These are but a few examples of the possible problems that blur the supposedly clear-cut distinctions.

It is easy to say that problems such as those outlined in the preceding paragraph can be solved by objective analysis: the determination of organizational responsibility and the institution of firm measures of correction. Unfortunately, sometimes we are dealing with situations in which it is difficult to get basic agreement on one or all of these steps in the general solution of the problem. The line people are production oriented and tend to be impatient with anything or anybody whose efforts seem to them not to be helpful in discharging their basic responsibility to "get out the work." The staff people are technically oriented and tend to be impatient with anything or anybody whose efforts seem to them not to be helpful in meeting carefully formulated operating methods or procedures designed to take advantage of new technology. The ironic part is that both groups usually have pride in their work, both groups agree on the need for success, and both are trying hard to meet objectives. Unfortunately, both groups may also have strong convictions; with this, each may have the ability to make the other "look bad." The trick is to have them work together; this is the second reason for discussing the line-staff relationship. The institution of a formal cost improvement program helps in this, because the line has the new assignment of meeting goals, and the staff has the incentive of receiving credit as improvements are made. Yet neither of these groups can do the entire job by itself, and a successful joint effort is necessary before either can succeed.

Problems in Industrial Engineering, Operations Research, and Information Systems

The second area of line need for technical assistance from the staff is that commonly associated with the disciplines of industrial engineering, operations research, or information systems. Distinctions among these disciplines sometimes become a matter of semantics, but generally speaking they are concerned with the definition of projects and the collection and analysis of data relating to these, the formulation of mathematical models of the problem area, and the analysis and design of information systems. In addition, industrial engineering usually is responsible for labor standards (which may have to be adjusted) and the information systems people for reporting and control procedures. It is an unfortunate fact that the relationship between these particular staff people and the line managers and supervisors is often a very unsatisfactory one. This is true because the industrial engineer sets time standards and changes methods, and both of these activities may in the past have been a source of controversy. After all, if the standards are "loose," the job of line supervision is in some respects easier. And methods changes require .new patterns of behavior, which can be troublesome to install.

As an example, let us suppose that there is no cost improvement program in effect. In too many cases the burden of improvement falls on the industrial engineer. Let us say that this is the case, and the industrial engineer, who also works in a project format, approaches a line foreman, saying, in effect, "Joe, this preinspection operation you have with the four new people was put in last year because we had trouble with the vendor. Now that seems to be cleared up, and I am here to look it over again to see if we still need the four people."

Joe immediately scents trouble. He may know that the operation is now overmanned, and could be done by one man rather than four, but it is convenient for him to have the extra manpower; besides, how will he break the news to the shop? So he will probably try to evade the issue or at least put it off. "I'd like to help you out," he will say (what he really means is "Which way did you come in?"), "but while these men do not look busy now, there are many days on which production could not proceed without all of them. I still have problems with the material. Besides, this is only a small operation; why don't you look in the assembly area where the real money is spent? I just can't spare anyone. And I wouldn't want to lose production because I was putting bad stuff into the process."

Now the industrial engineer has a problem, because he is on notice that Joe is going to be difficult, and also that the spectre of lower quality because of tighter manning has been raised. He may still pursue the matter and carry it through to a cost improvement, but he realizes that this will not be easy and certainly will take extra time. The basic problem here is that the industrial engineer represents a threat to a comfortable shop practice. The line supervisor also feels no compulsion to cooperate in making his own job more difficult, particularly when he cannot see where he himself will benefit in any way.

Let us replay the same situation, but this time with the condition that a cost improvement program is in force. The first reaction of the foreman will probably be about the same. There is no disguising the fact that reducing man-power will make it somewhat more difficult for him to operate. But he will now have additional thoughts, which may run somewhat as follows: "The boss seems to be really serious about this cost improvement program, and I know that I can do without some of the men. The personnel department says that they still have plant job security, so they will not suffer. And if I make them available for transfer, when they move they will be off my budget and I will have met my cost improvement goals for this quarter and also the next. Besides, everyone knows that as a supervisor I am under constant pressure to meet these cost goals, so the men will not blame me for their transfer, but will see it as my response to this pressure. So I'll go along with the industrial engineer and help him. In fact, I guess I should have thought of it myself." Under these circum-stances the change is much more likely to be made, and to be made with a mini-mum of fuss. This may not seem to be a very forceful example, but in fact the situation outlined is quite typical; the motivation given the foreman by the cost improvement program is the key factor. For a good foreman has both formal and informal control of his operation, and can use many stratagems to facilitate change or to impede change. We want him to feel that the former is in his best interest. Appeals to companywide goals, which he regards as somewhat nebulous, are not as effective as creating a situation in which he can see himself as directly involved and can see that he will benefit personally from improved operation.

Changes in Equipment and Facilities

The next area in which line and staff must work together is in the design, installation, and maintenance of physical facilities. Many cost improvement projects involve new equipment or plant alterations. The line managers eventu-ally must operate these, but it is not their responsibility to design and install the facilities or specify the equipment. The equipment is usually the responsibility of the process group, already discussed. Facilities design, affecting plant expan-sion or alteration, is in most cases made by plant engineering.

For example, in one plant a new storeroom was built adjacent to the pro-duction area, and a conveyor system installed to improve material supply and control. The line supervisors were going to have to make this work, yet plant engineering was responsible for design, construction, and installation of equip-ment. The need for close cooperation here is obvious. The motivation for this change was the cost improvement envisioned through more rational material control and physical security.

This type of project is a natural for creating conflict, and thus benefits from the community effort of a cost improvement program. The same is true of all process changes that involve new equipment. The ultimate effectiveness of such changes depends on the motivation of both line and staff. If each group receives credit for the joint accomplishment, a very human tendency to overstate individual contributions is lessened. A cost improvement program is of course not a necessity; after all, new equipment has been and will continue to be installed in factories. But many problems encountered in this activity hinge on the spirit of cooperation among the various people responsible. Cost improvement programs have an impact here, because of the obvious advantage of an appeal to the self-interest of each group and, at the same time, the requirement that the entire project be successful before this self-interest can be satisfied.

In addition to the line-staff cooperation necessary in the installation of new equipment and facilities, there is also an opportunity for cost improvement in the area of maintenance of plant and facilities. In many cases plant maintenance is under the same supervision as is plant and equipment installation. Here there is an obvious area of cost improvement in that the line goals of improved output and quality may depend upon the state of adjustment and reliability of the equipment, which is usually a responsibility of the maintenance department.

For example, in a company that had automatic packaging equipment running at high speeds, production goals were set based on the maintenance of these speeds. The earnings of the operators depended on this, which of course made it a sensitive matter. Yet the staff maintenance mechanics who serviced the equipment were faced with the problem that the operators were making unauthorized adjustments which increased output temporarily, but led to excessive downtime. This condition had persisted for years, with each group blaming the other for the problem. Finally, a goal was set under a cost improvement program to work toward almost uninterrupted operation. Both groups responded. Some might say that this should have happened with or without the program, but the savings were real dollars, and a long-standing loss situation was corrected.

The point is that the line needs help in getting into operation and continuing to operate new equipment and facilities, and the staff needs line cooperation in achieving technical progress. A positive effort to make it easy for both to cooperate is an obvious necessity, and in many cases a cost improvement program is the answer.

Personnel Problems

Many cost improvement projects result in changes in job assignments and manning levels, or in a demand for new skills in the people involved in the change. In addition, very natural questions arise concerning job security, seniority, compensation, and job evaluation. In this area the line managers need staff assistance from the personnel function. For example, if new equipment is being installed with the expectation of a net cost improvement through reduction in direct labor, the impact of the training costs and job reclassification must be considered.

In a specific example, the New York City Police Department management proposed changes in the shift coverage of patrolmen, which were supposed to result in substantial cost savings. The patrolmen's bargaining unit agreed to the changes, but only if premium pay were to be given to patrolmen working certain shifts. Substantial savings still were realized, but not of the magnitude originally estimated; the difference went to the patrolmen as new, extra pay. (This is a significant example not only because it shows the importance of employee relations, but also because it involves public employees.)

The fact is that the technical aspects of personnel administration can be quite complex, and certainly we cannot expect line personnel to become really expert in the details. If line people are "firm but fair" and have a working appreciation of the union contract, we usually are satisfied. But part of every cost improvement project that involves people is a financial analysis of the personnel costs and a statement of personnel changes. The personnel department is automatically involved in such changes, and a satisfactory line-staff relationship must be developed. Although the relationship may not have a direct connection to the origination of many joint cost improvement projects, personnel usually has to provide staff assistance in the form of training, transfer, job classification, and wage administration.

Accounting and Financing Problems

Finally, operation of a cost improvement program absolutely requires the development of real understanding between the line organization and the accounting function. The obvious essence of a cost improvement program is an evaluation of results in financial terms. Thus one function of the accounting department is to act as scorekeeper. In addition, each supervisor may be required by the program to analyze his own expense budget; here again he needs the technical assistance of the accounting department.

It may seem that the need is entirely on the part of line personnel, but this is not so. In the day-to-day operation, exclusive of any cost improvement program, the accounting department depends upon line management to see to it that the original reporting of hours, material, and other aspects of shop operation is accurate and timely. A very common complaint of the accountant is that shop reporting is not done well. The complaint is frequently valid, unfortunately, and is by no means confined to the shop. Record keeping at the source is a problem in maintenance, material control, customer service, and elsewhere. Errors in initial reporting are particularly troublesome when a computer is involved, which is almost everywhere today. So a side effect of the providing of

staff assistance to the line by the accounting department should be the development of an improved working relationship between line and staff. The staff
responsibility for accounting practice is not new, of course, but the spirit of
cooperation can be improved within a cost improvement program.

ESTABLISHING THE LINE-STAFF RELATIONSHIP

Mechanisms for establishing a good line-staff relationship vary widely.
Some companies transfer people between the line and staff as part of individual,
personal development programs. Others conduct formal training and appreciation programs on the subject of organizational responsibilities. But technical
competence requires time and work, so we cannot expect anyone to master
several technical areas; the real expert usually feels that he scarcely has time to
stay abreast of his own field. To take advantage of the expertise available, we
must be sure that every incentive is offered to obtain a free flow of opinion and
a spirit of cooperation. As a general statement, two broad concepts seem to be
useful here, and can be strengthened through a cost improvement program:

1 / The focus for effort that is offered by the formal structure of the
program and by the project approach takes the broad objective of cost
improvement and divides it into manageable pieces. Furthermore, the
project allows us to organize the line and staff people into teams of a
few people. The man who has damned "the accountants" for years
finds that the fellow with whom he now is working toward a specific,
agreed-upon goal may not be such a bad fellow after all, even though
he is "an accountant."

2 / The sharing of credit, which is part of the cost improvement philosophy, makes it possible to recognize the contributions of each member
of the project team. The topic of joint participation has been discussed earlier in this work. All that needs to be brought out now is
that in many organizations the line-staff relationships are in disarray.
But with strong top management support, a cost improvement program provides a face-saving device for renewal of more constructive
attitudes and inhibits the very human but sometimes destructive self-
interest we may now observe.

One other observation concerning the role of the staff is in order. It is true
that a basic staff problem is in extending their usefulness by serving as a catalyst
for improvement and as a staff resource to that line operation. But this is by no
means an indication that the staff and top management should operate only in
this way. In fact, a good staff will provide top management with analyses of
trouble spots, appraisals of performance and proposals for improvement. This
should go without saying. The expression "completed staff work" is a reflection
of this part of the staff's usefulness. The point which has been made in the

previous discussion, however, is that in addition to the internal work of staff departments, working with top management, there is a need in the conduct of cost improvement programs for the staff to work closely and provide technical support to the lined organizations in projects not primarily selected by the staff. In brief, the staff is accustomed to working upon projects developed through its own initiative. In cost improvement work, it must develop the knack of working with the line on projects developed through line initiative.

7

Setting Specific Goals

It is an axiom of human behavior that people work most effectively when they are striving toward a realistic well-defined goal. This is true for a number of reasons, but certainly foremost among these is the security of the feeling that the dimensions of the task are known, and that plans may be formulated and commitments made on the assumption that meeting the goal set will be a signal of success. There should therefore be no feeling at any level of the organization that the setting of goals is a unilateral and arbitrary exercise of top management pressure. Instead, the setting of goals should be an exercise in participatory management, with a great deal of input from several levels of the organization. This is true because not only must the goal be fair, but it also must be seen to be fair.

Working toward a goal is not only good human relations, it is an essential part in the management process. How could we plan and schedule without goals? We have said that the cost improvement program deserves serious consideration partly because it gives the manager "something to manage." And the essential form of the management process consists of setting goals, planning, scheduling, execution, and follow-up. So the need for goals is obvious. Goals do put on pressure for achievement, of course. That is why we set them. But with so much at stake, it is only fair to all that the goals be rational, representative of something attainable, and universally and equitably applied.

HOW SHOULD GOALS BE SET, AND BY WHOM?

The essence of the setting of goals is that the process be perceived as fair and reasonable by those to whom the goals will be applied. If it were possible to use some magic universal formula, there would be no problem. The closest we can come to this is the flat-percentage-cut philosophy already discussed. And although this method has the advantage of simplicity, it has disadvantages that outweigh this.

The setting of specific goals should not be done until a preliminary analysis has been made. If we think about it, this apparent inversion of the steps in the scientific method makes sense. It has been pointed out that business problems are recognized *as* problems because some number is not consistent with what our experience has led us to expect. Scrap is too high or orders received too low, for example. Furthermore, opportunities for cost improvement are governed by the amount of cost involved and by its distribution. We should consider these factors, department by department, before we set specific goals.

Initial Input-Output Analysis

The mechanics of a cost improvement program include analyses appropriate for the setting of goals. Basically, there is an overall two-step input-output evaluation:

1 / A Pareto analysis of costs; that is, a listing in order of magnitude of the expenditures (budgeted and actual) for each department. This gives the input side of the relationship.

2 / A listing of the products, services, or work units of output for each department, against which the costs are distributed. This describes the output.

It is important that we give our attention to both parts of this relationship. Too often the cost or input is all that is considered. This is, however, a restrictive and sometimes self-defeating practice, because it does not systematically bring up for examination the purpose of each expenditure, and it tends to assume that all operations can show substantial improvement simply by better methods and management. There is truth in this, but the best cost improvement for any operation is to eliminate it completely; this requires a very broad view of the entire series of operations. To be certain that this view is taken, we need a systematic analysis with no holds barred. An input-output examination of all phases of operation serves this purpose.

This initial analysis is done by all line supervisors and managers. Usually the form of participation involves group meetings at which each supervisor brings in the budget and work-unit analysis of the activity for which he is responsible. The budget part is easy, although it does have the effect of establishing the line-staff relationship with the accountant a little more clearly. But usually there is no problem in simply arranging the expenditures as shown in the regular chart of accounts in descending order of magnitude. Questions may arise about the logic and procedures in accounting, but this is more a problem of education than of conflict.

The definition of output usually raises the most questions. First, the concept is sometimes new, and, second, particularly in indirect-type activity, the units of output are sometimes difficult to define. But the very act of reducing both parts of the input-output relationship to manageable, understandable form is a big first step. At the very least, each supervisor will see that his operation has a structure which can be analyzed with a view toward improvement. Furthermore, the supervisor may realize that, although he controls the internal operation, the information which describes this operation exists outside his office. This information is of course available to others who may use it to appraise his work. The implication is clear that he should himself use the data in his own cost improvement work. In any case, the setting of goals becomes more realistic when we have made the initial input-output analysis, because we now know the nature of the costs as well as the nature of the demands that generate these costs.

Sometimes goals are set directly from the first broad analysis. Usually a discussion of individual items will bring forth ideas for improvement. As the costs are examined systematically, the assumptions governing the form and character of the output behind the costs will be put to the question. This is healthy, and certainly adds dimensions to the possible improvement. At the same time, this process will raise questions of specific policies, and a decision may be made to resolve these before setting goals.

For example, a maintenance supervisor may say that he could probably improve his operation by better planning to reduce travel and waiting times, but he may question whether or not there is enough potential improvement to make a planning system worthwhile. He could answer that question with a work-sampling study. Or a production supervisor might remark that the small orders he now is required to fill are costly and might better be consolidated, even though this might increase inventory carrying charges. It will take investigation and perhaps a new policy to resolve this. The point is that these questions were raised by the broad initial analysis, but that we need further analysis to appraise the extent of improvement which may be incorporated into a reasonable goal.

One approach here is to have those departments in which personnel costs are a major factor take a work sampling of the people. This is usually done for a period of a calendar or accounting month. At the same time, these departments will pay particular attention to their records of units of output, which are often kept for the same time period. Matching these data will give us the input-output data we need. Departments in which materials or facilities charges are major items of cost may not take a work sampling, but will make analyses of the product (as in value analysis), the inventory policy or distribution, and sales practices as they affect the operation. At the end of this more detailed analysis period, it is reasonable to set specific goals. In addition, this analysis activity will have had the effect of publicizing the cost improvement effort and probably will have brought out some sensible ideas for avenues to accomplish this.

Setting Goals: The Cost Improvement Committee

The setting of goals usually is most effectively done by a strong cost improvement committee. This committee should include the top executive in the particular organization (probably the president or general manager), his assistant, the director of cost improvement (who may serve as secretary), and the head of each major function (such as sales and manufacturing). The committee meets to set goals and to review progress. Needless to say, the time of these people is valuable, and we would expect others in their organizations to prepare the material that the top person will bring to the committee. But the actual determination of goals should be done by a cost improvement committee of the nature described. Their meetings should be scheduled, and full attendance should be expected.

There are several basic reasons for the formation of yet another committee, which will demand the time of busy people:

1 / To provide extremely visible evidence of top management's interest and support.

2 / Goals are much more likely to be regarded as fair by individual departments, since the top executive from each department will have participated in setting the goal. In brief, he will take the message back to his people in the form, "I have agreed to this based on what you tell me," rather than, "This is what they gave us" (whoever "they" is).

3 / A firm basis for cooperation at all levels is established when those at the highest level are seen to be working together.

4 / With all major departments represented by the top people, there is more apt to be a realistic atmosphere and less likelihood of circular reasoning. For example, the production executive may say to the president in private conversation that he cannot improve his costs much until the sales manager stops making unreasonable promises of delivery to customers. He is less likely to say this in a group meeting unless the facts support him absolutely. Even then, we at least have the problem out in the open for discussion.

5 / Some companywide information may be considered confidential within the company, so that normally only a group at this organizational level would have access to it.

The manner in which goals are set will vary, but the usual practice is to bring in for each department the results of the preliminary input-output analysis, together with an estimate of the amount of cost improvement that may seem to the department to be reasonable, the assumptions that underlie the estimates, and a schedule. The assumptions should include the broad needs for interdepartmental cooperation. For example, the sales manager might say, "Assuming that our forecast is accurate to within five percent, and that manufacturing can maintain its projected production schedule, we can improve our distribution costs by the amount stated." This type of statement may seem to be nothing more than a method of building an escape route in the event that things go awry; but it should be remembered that this is a two-way street, and above all that the president is listening. So the combination of a knowledgeable and critical audience plus the motivation to satisfy top management's expressed determination to obtain real cost improvement usually results in a reasonable set of goals.

These goals usually are expressed in dollars and are issued with a time schedule for their realization. It is the job of the cost improvement committee to reach agreement on the goals, with the understanding that each member is responsible for his own. The goals should be attainable, yet should represent substantial cost improvement. No one likes to commit himself to a goal which is too high, for he knows that this will put unreasonable pressure on himself and his people. Yet both peer pressure and executive pressure work against setting a trivial objective. Fortunately, the initial analysis of input-output data provides a common base for discussion, presented in a manner which is consistent and that uses accounting language familiar to all.

For example, the manager of a shop that fabricated welded pressure vessels made an initial analysis of input through the budget. This revealed that about 20 percent of his cost was material, 40 percent direct labor, and 30 percent overhead charges allocated by labor hours. Some of his other expenses were so closely related to these accounts that they were included there, and some were not examined because they were small. An analysis of the output units showed that all the vessels were designed to meet very rigid code requirements when in fact only about a quarter of the units sold really needed this level of quality. In the others, the customer got a real bargain. The reasoning had been that if more than one quality standard were to be manufactured in the shop, costs would not improve too much and the possibility of error would be great. The manager raised the question with the sales and engineering staff, and was surprised to find a very receptive attitude toward manufacturing at the appropriate quality level, and not always seeking the highest level. In fact, sales and engineering had assumed that having one level was best for the shop. If cost were no object, this might be true and the matter dropped. But now cost was a critical item, and an immediate review of the policy was started.

Personnel costs were analyzed with a work sampling study (Table 7-1) and showed promise of some short-term cost improvement.

Table 7-1.   WORK SAMPLING RESULTS—
              FABRICATION SHOP

| | Foreman | | | | | | |
| | A | B | C | D | E | F | Total |
	%	%	%	%	%	%	%
1. Fitting	36	43	30	23	27	37	36
2. Weld (Arc)	13	22	17	22	33	27	20
3. Machine Set-Up	3	1	7	2	6	2	3
4. Grind./Chipp.	12	14	18	27	10	17	15
5. Layout	4	3	3	--	5	--	3
6. Cleaning	9	--	7	3	10	6	5
7. Talking	6	2	3	2	1	--	3
8. Walking	6	1	3	2	2	1	4
9. Print Reading	1	1	2	--	1	--	2
10. No Contact	3	5	5	11	2	3	4
11. Other	7	4	5	9	1	7	5
Total %	100	100	100	100	100	100	100
Number of Observations	1058	1513	671	243	260	626	4371

	Number	%
1. Comb. Welder	32	41%
2. Arc Welder	7	9%
3. Fit/Up	31	40%
4. Assy/Fit-Up	8	10%
Number of Men	78	100%

For example, only very limited use was being made of automatic welding machines, which weld faster and in most cases more uniformly than an operator. So something could be gained by systematically working to increase machine utilization; this is within the authority of the supervisor. Furthermore, half the shop personnel were in the high-cost wage rate of certified welder. Only about one fifth of the work utilized this skill. Of course, some flexibility is necessary, but the price for this seemed too high. In the course of his investigation of high welding cost, the manager came across a time card which showed that still another employee had taken tests and training to become a welder, at a higher wage, on a Sunday while being paid double time. In brief, at the meeting of the cost improvement committee, the manager had a pretty solid idea of the improvement he could go after and also of how to measure his results. In addition, part of his overall program was a cooperative effort with design, purchasing, and engineering.

The reader should understand that this business of setting goals is the key step in the mechanics of operating a cost improvement program. If the goals are fair and substantial, top management will be able to support the program and be quite positive in its administration. If the goals are universally perceived as being fair, the program can be made to be a very powerful force for change. Management can in good conscience insist upon a strong effort to meet goals. If the goals are substantial, it will be obvious that the effort and the disruption of routine are indeed worthwhile. The setting of cost improvement goals and the pressure for their fulfillment are among the duties of the cost improvement committee; this is made up of the most powerful men in the organization and should have the capacity for leadership.

Of course, no company or organization is immune from self-interest and the workings of internal politics. The cost improvement committee brings together people who hold strong and honest convictions, and some strong and honest differences of opinion are bound to emerge. But the competent top executive realizes this, and usually has the skill to place these wholly natural differences in the framework of advocacy rather than of conflict. And if the basic budget and work-unit analyses have been done conscientiously, everyone will find enough challenge to keep him busy, without worrying about his peers' performance. The good executive will understand that each department and each manager is different. These differences are natural, and as long as the common objective of meeting cost improvement goals is kept in mind, no harm is done. Indeed, differences in point of view should be actively sought. Advocacy and the competition of ideas are almost essential to progress in any case. To have open discussion among peers is far more desirable than to have evasion and deprecation of goals set by executive fiat.

Department Cost Improvement Committee

We have said that in preparing his presentation of suggested goals to the cost improvement committee, each manager probably has already involved his organization in the initial input-output analysis and the critical review that follows this as a matter of course. So once the goals have been set by the committee, the top manager in each department sits down with the people who have done the preliminary work. He discusses with them the actual goals set and explains the reasoning of the committee, particularly in those instances when departmental proposals have been substantially changed. In effect, this group acts within a department as a department cost improvement committee and works to meet the several department project goals, which, taken together, make up the overall department goal. The techniques used to accomplish cost improvement will be discussed later. But this discussion of the organizational structure is necessary in that the actual technical work should be done within a good framework of management action. The process of setting goals is an integral part of this framework.

ADMINISTERING THE PROGRAM

The usual arrangement for administering the program is to set up projects for cost improvement based on the accounting and work-unit control systems. Then line and staff personnel are assigned to teams that are responsible for each project. Department management monitors progress, by project, and should provide support as necessary. There should be periodic internal review meetings and summary reports to the scheduled meetings of the cost improvement committee. No particular discussion seems necessary concerning this, because it should follow the usual management pattern of the company. In fact, the central idea here is that cost improvement be treated within the management pattern which is familiar to the employees and is being used by management in conducting their everyday business. If the appropriate management effort is made, the program can be expected to produce results. As a word of caution, the timing of some of the projects can be expected to slip, mostly because the concept of planned and scheduled change is new. But experience has shown that once the familiar management proce' 'f setting objectives, planning, scheduling, execution, and follow-up has been established, administration of the program is fairly straightforward. The setting of goals is the first step in this process.

As progress is reported, the question will arise of how "firm" a project must be in order to include the dollar figures in any summary of results. It is a good idea to evaluate projects at three stages of their life and to keep these separate: (1) The first evaluation should be an "estimated" figure, given as each project is started. This will be based on whatever input-output analysis work has been done. (2) When the project is completed and installed and the accounting data become available, a "claimed" figure should be entered against the project. This should be fairly firm, of course, but represents only limited experience under the change. (3) After a period of time of two or three months, an examination can be made of actual experience, and an "audited" improvement can be entered. This will be discussed in the next chapter, but the term is almost self-explanatory.

Use of this three-stage system of reporting cost improvement leads to greater precision of understanding. It is quite clear that "audited" figures are representative of real accomplishment and that "estimated" figures are not firm. At the same time, the relative state of progress is evident. Furthermore, this system allows for changes that may occur as the project develops. Above all, such a careful evaluation of projects gives the cost improvement program credibility. And even though some of the projects may not meet their original expectations, it is important that any stated improvement be genuine. This should also be reflected in the regular cost accounting reports. The program will gain more acceptance through honest reporting of modest gains than through optimistic overstatement.

8

Conducting Systematic Audits

In Chapter 7 a procedure was suggested for reporting the dollar amounts of cost improvements in three stages. The first, termed "estimated," is entered in the system as a project is started. The second, entered when the project has been completed and installed, is given the designation of "claimed." Finally, each project should be reviewed formally after a period of from two to three months and an "audited" amount established. This process of careful review is the subject of this chapter.

There are several reasons for an audit of cost improvement projects, done as a matter of course and established as part of the regular procedure, the most important of which are as follows:

1 / The general premise that any numerical measure used in financial reporting and particularly in performance appraisal should be capable of audit.

2 / The fact that an audit will be made as a matter of routine will discourage overstatement of expected improvement.

3 / Conducting such an audit and then analyzing the results is a most useful device for uncovering both reasons for failure to attain expected results and opportunities for further improvement.

4 / Such an audit, made after two or three months' experience, gives

some assurance that the improvement has been given a chance to become established over time, and has had a fair chance to prove itself.

5 / Such an audit indicates that management is following up on its commitment to cost improvement.

6 / The audited figures represent real, honest improvement, and can be defended as such.

Before taking these reasons in order, it is appropriate to discuss the audit procedures. The period of time between the installation of a cost improvement project, with submission of a "claimed" dollar amount, and the performance of the audit should be at least two and probably three months. This time period is suggested because of the following:

1 / There is a learning process involved, and it will take some time simply to learn new procedures or methods.

2 / The old methods or procedures must be phased out; this sometimes results in problems such as using up materials or running parallel systems for a while.

3 / If personnel savings are made, it may not be possible to transfer people right away, and they may have to be carried by the department for a short period. As a rule, no savings should be allowed until force levels are actually reduced or new work added.

4 / Usually, it takes a full accounting period or two to work out the old procedures and have the new ones reflected in the accounts.

PERFORMING THE AUDIT

In performing the audit, the regularly established accounting system should be used. Part of the initial project statement should be an analysis of the projected improvement in terms of specific accounts from the regular chart of accounts. For example, a particular project may envision a reduction in direct labor costs that would more than offset a concomitant increase in direct material costs. The appropriate accounts should be identified beforehand, and it should be agreed that the regular reporting system would in fact reflect the changes which are planned. Furthermore, the accounting practices by which overhead is applied and labor hours and material costs are charged should be carefully reviewed. This is done so that the impact of the particular cost improvement project may be isolated, and the project evaluated in commonly used management reports.

The reason for this insistence on the reduction of cost improvements to the usual format of existing financial reports is that it is these reports which are normally used to measure operating effectiveness; management should take the position that, if improvement in operation really has occurred, this improvement should be measured and reported in the same way that the operations themselves are usually reported. In other words, management has for years used the conventional reporting system—usually a profit and loss statement—as the basic measure of financial performance. Managers are familiar with this system, and have their own ideas of the meaning and importance of each part of these financial reports. Individual managers usually have made use of trends and ratios among the various items to spot trouble and to appraise progress. In addition, the existing format has been used to represent operating conditions to the board of directors and the stockholders. It would be difficult to make the case that a new system or some sort of special reporting method is necessary for a cost improvement program. The obvious reaction to this is to raise the question of the adequacy of the established system. This is not our intent.

BASIC REASONS FOR AUDITING

The first basic reason for conducting an audit is the general philosophy that every financial measure should be capable of audit. On the positive side, whenever management uses progress in cost improvement as one of the measures of supervisory performance, it should, like all such measures, be accurate and meaningful. Top management itself should have an interest in the credibility of performance reporting, so that they can appraise the effort required of them in terms of solid results. Finally, we all have seen cases in which the initial reaction to an appraisal that is unfavorable is to attack the validity of the numbers on which the appraisal is based.

The second reason for conducting an audit is that the very fact that an audit will be made will discourage overstatement of cost improvements in either of the two preceding phases. We are less likely to see inflated numbers as "estimated" or "claimed" improvements when it is established practice to run an audit of all projects. This practice also serves as protection against a strong manager simply making the statement that an improvement has been made, and asserting that this discharges his obligation under the cost improvement program. Instead of questioning such a claim, we merely enter it as "estimated" or "claimed" improvement and wait for the "audited" results. And no one can say that he is being singled out, or that his appraisal of his own department is suspect. The audit is made of every project, just as the public accounting firm conducts an audit of the institution itself, and for some of the same reasons.

The third reason we audit cost improvement projects is a more positive one. By conducting an audit and then analyzing the variations between "claimed" and "audited" figures, we lay the foundations for further action. If the "audited" figure is substantially less than the "claimed," we should find out why. For example, one cost improvement project was installed that depended for its success upon observance of a strict "no overtime" policy. This policy had been adhered to at installation, but after two months' time had been allowed to loosen. The audit showed this, and the "no overtime" policy was reaffirmed.

The audit will also uncover areas in which the "audited" improvement was greater than the "claimed." These cases are usually indicative of alert supervision and should be pursued aggressively. For example, the manufacturer of a wide variety of power hoists had instituted a product simplification program as part of cost improvement. This reduced the number of models from 37 to 8, and was expected to reduce manufacturing and inventory costs. The savings were entered as both "estimated" and "claimed." When the audit was conducted, it was found that setup costs for the change had been overestimated, and that cost improvement was thus understated, which led to a review of scheduling practice and subsequent additional improvement. This might not have been discovered without the audit.

The procedure of audit and analysis of variation also gives insight with respect to the general types of cost improvement projects that succeed in a given situation and those which seem to fall short. Such information is valuable to the manager. For example, in one plant cost improvements involving materials usually were successful, and those involving the processes of manufacture seldom met expectations. The manager investigated this and found that the procedure for instituting process changes was cumbersome, that some foremen were not even aware of the new adjustments to equipment, and that some instrumentation was inadequate. On the other hand, materials changes had to be documented and involved purchase specifications, which gave a more positive response to change.

Another basic reason for the auditing procedure suggested is simply that any change will almost have to remain in force for the two- or three-month audit period. This not only gives everyone a forced opportunity to learn how to use the new method or new material, but also works out the last vestige of the old. The effect of this is to establish the change if it is good, or to supply firm evidence of failure in the event a change does not live up to expectations. This last is important, because all projects do not succeed. Thus an audit not only provides a solid base for decision on the value of a change, but it also is a reasonable way of defining failure to achieve projected goals. In this way we protect ourselves from "throwing good money after bad." As a corollary, continuing management interest is demonstrated by the audit.

Finally, a careful review will be seen by all to give credibility to the final "audited" figures for cost improvement. Almost invariably, a cost improvement program will produce real and significant results. It is important that these results be accepted as genuine, because too much effort has been expended and too many people affected to have it any other way. There have been cases where results have been deliberately overstated, at considerable cost in employee morale. One manager of such a program felt that by being generous in evaluating improvement he was encouraging his people. It turned out, to his dismay, that the employees felt that he was "playing games," and they referred to the improvements as "Chinese dollars." We should depend upon the employees' common sense to support a serious effort. We do not need such gimmicks and should paint the results as they are, "warts and all."

In summary, a systematic audit performed after a reasonable period of time and using the framework of the established accounting system is a valuable part of any cost improvement program. Analysis of the results of such an audit will provide management with a valuable feedback on the realities of the operation. Systematic audits lend credibility to the entire effort. This is essential both to employee morale and to the evaluation of the program by management.

II

DEFINING THE PROBLEM FOR MANAGEMENT ACTION

In Part I we discussed the management framework within which cost improvement programs operate. In Part II we shall discuss the techniques used for the preliminary analysis that is made to define the problem for management action. The first step in any management effort is such problem definition. The initial analysis should be done in such a manner that managers and supervisors at all levels will be involved, that the analysis will direct our attention first toward the important aspects of each operation, and that the foundation will be laid for the improvement we seek.

The three basic concepts that pervade Part II are (1) the quantitative nature of problem statement, (2) the emphasis on Pareto's law, which almost forces us to concentrate on the "vital few" at the expense of the "trivial many," and (3) the emphasis on the input-output relationship in cost improvement, and the systematic appraisal of what we get as well as what we spend.

Techniques and procedures for the initial analysis discussed here are relatively simple and within the capabilities of the people who will be asked to use them. The initial analysis should be done throughout the entire organization. It is suggested that this be done before decisions are made about the specific areas in which concentrated effort will be expended. It is almost certain that this initial analysis will raise fundamental questions about current practices and, at the same time, provide insight to the direction improvement should take.

9

The Quantitative Nature of Problem Statements

Whenever we institute a cost improvement program, there is implicit in this action the notion that we have a problem in costs. This is a rather obvious statement, yet it is pertinent to the discussion of problem definition. Einstein and others are our authority for the concept that "a problem, once stated, is half solved." But how do we recognize a problem in the management of operations? Problems of personal life are many times of a qualitative nature, where intuition and emotion are governing factors. But problems of business or operations management are more likely to involve quantitative measures, and are in fact recognized *as* problems because some quantitative measure does not behave as our experience or expectations lead us to believe it will. Cost problems do not exist in the abstract; they are associated with numerical or quantitative measures. The initial analysis is made for the purpose first of establishing what these numerical measures are; it then becomes a matter of management judgment to determine quantitative goals for improvement, and of management action to achieve these goals.

ADVANTAGES OF INITIAL ECONOMIC ANALYSIS

One obvious advantage of a comprehensive preliminary analysis is that it will establish basic input-output cost relationships in all parts of the organization; for some parts, no such measure may have existed previously. For example, in indirect activities such as clerical and maintenance work, such measures as cost per invoice or cost per work order may not have been established. Not all the areas measured will be considered problems, of course. But whether a problem exists or not, we should know the input-output relationship. Even where we can expect only the improvement normally associated with the learning curve, our initial analysis will provide us with a "bench mark." Since at this point we do not really know which items will be affected by the cost improvement program, it is well to have as much pertinent quantitative information as possible as a basis for management evaluation and decision.

The preceding thought is worth further discussion. One underlying objective in management's support of cost improvement programs is to institute a searching self-examination in all areas of operation. Most of us have the feeling that cost improvement is possible perhaps for the other fellow, but that we are doing a good job. One reason is that we are caught up in a routine which we know to be fairly effective, if not fairly efficient. The temptation is to concentrate on day-to-day operations, regarding this as "our job." The job can get difficult, but if we are competent we can overcome the difficulties. We tend to become absorbed in the work, and not to question matters of established practice. In addition, we know the exact dimensions of our present job, and the element of personal security leads us to respect these.

To effect cost improvement, however, we must change. This not only is an implied criticism of the way things are now, but also involves unknown outcomes. These factors combine to produce the whipping boy, "human resistance to change," about which so much has been heard. However, if we are forced to define our job in terms of cost input and work output, we automatically institute the self-examination that has been mentioned as one of management's objectives. The saving factor in this situation is that, if the job has not recently been subjected to objective review, it probably can be improved substantially.

Importance of Initial Costs

The importance of establishing the initial conditions as a bench mark is evident when we understand that cost improvement is a relative measure. We may find, for example, that it costs $300 to prepare a bid, or $12 to fix a leaky faucet, or 30 cents to paint a breadbox. We may have some idea of what our competitors are doing, but differences in quality, design, wage rates, or distribution costs may make comparisons difficult. It is therefore difficult to say what our costs "should" be. We do know very well, however, in which direction lies improvement, and we now know our point of comparison—the present cost. If we reduce these costs per unit or improve the quality without raising cost, we are progressing in the right direction. Very simply, it is the difference between what costs are at the start of a project and what they are upon completion—

always considering the output—that we take as the improvement. This is obviously a relative measure.

Finally, if we go after improvement over the initial cost, we may be seen to be declaring a form of amnesty for previous unsatisfactory practices. This may disturb some managers, but the realistic view is that if we are improving on our past experience it serves no constructive purpose to rehash yesterday's mistakes. After all, conditions were different, personnel were different, and we did manage to survive and even make a profit. Our problem is now; we want to do better in the future. That is where our interest should lie.

One of the aspects of any cost improvement program usually involves previous efforts toward cost improvement within the company. All too often these efforts are centered around some technique, such as work simplification, value analysis or zero defects. In many cases, such efforts have been effective in the short term, but have eventually been abandoned as those involved perceived that a universal effort was not being made in all areas. Cost improvement programs do in fact involve techniques such as those mentioned. But the organizational underpinning provided by a continuous cost improvement program simply was not present. If the perception is that from now on cost improvement will be a fundamental management concern, the various techniques will tend to be regarded properly as means to an end, and not the end itself.

Another very practical reason for obtaining a thorough initial input-output analysis is that we want no questions in the future concerning the actual results. People tend to have selective amnesia when memories are painful. To be consistent in our appraisals of success or failure against a relative measure, we must have well-established initial conditions.

Defining Work Units

The initial analysis or statement of the problem must include both input and output. This has been discussed previously and will be elaborated upon in the future. But in terms of any statement of the problem, both are required. This concept is particularly necessary in the intermediate steps of the production of goods or services and in the indirect departments in manufacturing, where such measures may not now exist. We first must define both the cost and the initial characterization of the work units of output. It has been the experience of the author that it is the latter which is most productive in instituting change, because this definition of work unit is not easy. However, it usually is most rewarding. For example, if it costs us $300 to prepare a bid, and we wish to improve on this, we should not simply concentrate our efforts on doing the same number of bids in the same form but at lower cost. This approach will undoubtedly result in some improvement, of course. But if we include in our statement of the problem formal considerations of whether the present form and number are appropriate, or whether we need the entire procedure, or even whether it is worthwhile preparing bids at all, we have expanded our opportunities for improvement. We may not make radical changes, but we should at least consider them. The initial analysis should point to this.

USING PARETO'S LAW

The general approach to most economic analyses done for management is to apply Pareto's law to the data. This concept has been around for centuries, but is best known in the formulation given it by an Italian economist named Pareto, who wrote early in the twentieth century. He observed that, in considering the multiple items in many classes of economically oriented populations, a relatively large amount of the money involved was concentrated in a relatively small proportion of the items that made up the population. Conversely, the relatively small proportion of the money that remained was distributed rather thinly over the relatively large proportion of remaining items. Pareto's law has many common applications. The "Fortune 500" list of corporations in the United States constitutes less than 1 percent of all corporations, yet accounts for a substantial fraction of the country's sales and assets. The "ABC" inventory system is another example, and is in common use because it reflects the fact that, as a rule of thumb, 80 percent of the dollar value of most classes of inventory is concentrated in 20 percent of the items. As another illustration, in one manufacturing plant 75 percent of the direct-labor hours were covered by 15 percent of the time standards. And an analysis of maintenance department work orders showed that 67 percent of the craftsmen's hours were charged to 15 percent of the work orders. The reader is encouraged to test this concept for himself.

The value to us of Pareto's law is that it provides a simple and understandable approach to our initial analysis or statement of the problem. The first step in any cost or input analysis is to have each manager or supervisor make up a list that shows, in descending order of importance, the costs associated with his own organizational responsibilities. For example, one electrical distributor's analysis of his operating expenses took this form:

Dollars	Operating Expense	% of Total Dollars
$180,300	Wages and salaries	55.8
69,200	Selling expense	21.4
36,000	Fixed expense (taxes, ins., etc.)	11.1
14,400	Supplies	4.5
9,800	Contract services	3.0
7,600	Telephone	2.3
4,100	Utilities	1.3
2,000	Miscellaneous	0.6
$323,400		100.0

The "miscellaneous" category included several separate accounts such as postage and petty cash. These data are sometimes presented in graphic form, as a cumulative curve, which has this appearance:

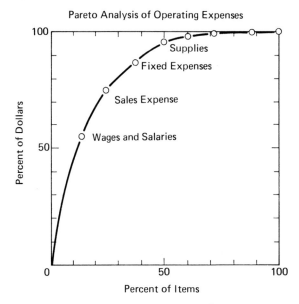

Pareto Analysis of Operating Expenses

(8 Items -- Each is Therefore $12\frac{1}{2}$% of Total)

Example. Total Expenses $323,400. Wages and Salaries Total $180,300. or 55% of the Entire Budget.

Figure 9-1.

As an example of Pareto's law applied in the initial analysis, let us look at the distribution of work-order size (in terms of man-hours) in a maintenance department that was involved in a cost improvement program. This analysis is an example of the flexibility of Pareto's law:

Hours per Work Order	*Number of Work Orders*	*% of Total Work Orders*	*Total Hours Charged*	*% of Total Hours Charged*
24.5 and over	58	4.7 } 15.1%	3,094.5	43.5 } 67.2%
8.5 to 24.0	129	10.4	1,687.0	23.7
4.5 to 8.0	149	12.0	894.0	12.6
2.5 to 4.0	195	15.7	631.0	8.9
1.5 to 2.0	250	20.1	440.5	6.2
0.5 to 1.0	460	37.1	367.5	5.2
	1,241		7,114.5	

The cumulative distribution gives this graphic presentation:

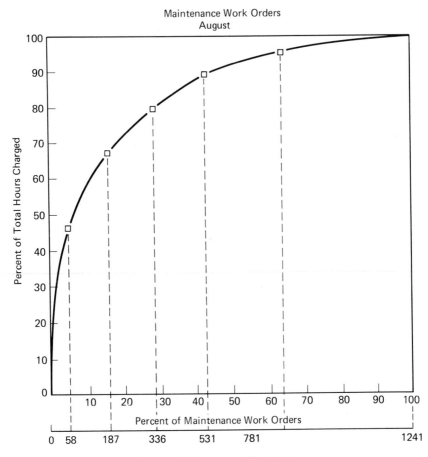

Figure 9-2.

The importance of such an analysis to management rests on two premises: (1) the time that can be devoted to cost improvement effort is limited, and (2) we must therefore concentrate this effort where the prospect of improvement is greatest.

A Pareto analysis of each department is in a sense self-directing in that the managers and supervisors who are themselves under many other pressures of time will tend to seek improvement in the larger items of cost. In addition, managers will have a systematic approach which probably will make sense to them. And top management will have a consistent form of problem statement, department by department, which is brought to them by the departments and which can be reconciled with regular accounting reports.

One important aspect of a Pareto analysis is that it usually brings out in a rather striking manner the fact that we should not devote the same amount of our effort to each item of output, but should first classify them in terms of their importance and then deal with them accordingly. For example, even though the format of each maintenance work order may be the same, we should not attempt to analyze, plan, and control the 37 percent of the work orders that account for only 5 percent of the labor cost in the same way that we analyze, plan, and control the 4.7 percent of the work orders to which are charged 43.5 percent of the labor cost. Any decision rules that are reasonable for one group probably do not make sense when applied to the other. Part of our problem, incidentally, may be that we treat them all the same. A simple and systematic application of Pareto's law to the basic inputs and outputs of each part of the organization usually is quite revealing, and in many cases guides us in the direction that further steps in the cost improvement program will take.

THE QUALITATIVE NATURE OF NEW OPPORTUNITIES

The Pareto analysis is, in the opinion of the author, absolutely fundamental to the initial analysis in cost improvement programs. Yet it is basically an analysis of the existing situation. A Pareto analysis provides a systematic approach to the attack on existing operations, whatever their nature. What the Pareto analysis does not provide in itself is a means of stimulating the exploration of opportunities in areas not covered by existing budgets and existing outputs of goods or services. This is not a "weakness" of the Pareto analysis technique, but rather a suggestion that while opportunities exist in the area of cost improvement and operations, other opportunities exist in changing or expanding the current scope of the business. An over-simplified version of this notion might be that cost improvement programs are designed to do better with what we have, but that other opportunities exist if we think in terms of what we might become.

The appeal of using an analysis different from the Pareto analysis is that we might possibly improve our profit picture by developing new sources of business and new products or services. This is, of course, how businesses grow. However, this text is limited to considerations of existing organizations. This does not mean that the development of new products or new businesses is not important; indeed, it is probably a more attractive road to profit. At the same time, there is more risk involved. There are texts which cover new product development and there is no intent here to suggest that this aspect is unimportant. We simply will not cover this topic.

A word also should be said about other techniques of improving the operations of a public or private enterprise. Many opportunities exist in the area of process and product research, organizational development and managing the capital structure of an enterprise. Again, these are attractive, and may possibly be included in cost improvement programs. But they tend to be long-term projects, and their effects somewhat difficult to evaluate. The author has chosen again to limit the treatment of this book to those areas which are pretty much under the control of the chief operating manager, and which in addition are relatively short-term in nature and which can be appraised fairly reliably. This limitation still leaves ample opportunity for cost improvement, of course. The most compelling argument in favor of the scope of this book is that experience has indicated that operating expenses of the type dealt with herein are those about which we can claim positive improvements made in a systematic fashion. This book is directed at managers at all levels who have responsibility for operating either a public or a private organization to meet production or service goals within firm budgets. This seems to the author to be ambitious enough in itself.

10

Analysis of Inputs:
Budgets and Costs

At this point it is appropriate to discuss the concept that any organization which has inputs of human effort and materials that it uses to produce outputs of products or services may be thought of as a system. The author is somewhat hesitant to introduce the term "system" because it has so many different applications and such widely varying special meanings. Also, it is unfortunate but true that the term has been applied over the past few years to situations which do not meet the basic conditions by which a system is identified. However, many reasonable applications of the term exist. For example, the interrelated clerical and computer activities that collect, manipulate, summarize, and feed back data are sometimes known as "information systems." In the field of electrical engineering, the term "system" is applied to the equipment, lines, and operating procedures and personnel used in the generation and distribution of electrical power. The systems concept is useful in understanding these and many other operational entities; the key points in any definition of a system are that several interrelated components operate to meet a given purpose, and that the salient features of this operation are inputs, outputs, a processor, feedback, and control. Although some of the mathematical models made of systems may be quite complex, the basic block diagram is quite simple and provides a good conceptual framework for our discussion.

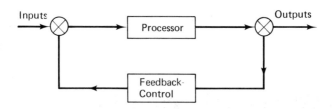

Figure 10-1.   Diagram of Basic System.

The rather elementary diagram given here is as far as we need go in our characterization of the system as such. We are interested in this because the elements of producing a product or providing a service usually form a system, and we should consider all parts of the system in looking for cost improvement. It is possible, of course, to construct much more complex models of an operating system, and these can be most valuable. In fact, if the system is complex and we are able to define it in mathematical terms, we do this as an aid to understanding the effect of change. However, our basic interest is in the fundamentals of cost improvement, and for this the basic concept is enough.

ANALYZING BUDGETS

The first step in our analysis, therefore, is to examine both the inputs and the outputs of our system. We shall defer consideration of the details of operation until we have established the relative importance of each part. The initial analysis of the inputs to the system should start with the operating, overhead, and materials budgets. This is logical, since cost is the common denominator of all inputs.

Budgets have two characteristics that make them attractive and useful to us in the initial analysis:

1 / Budgets are part of the regular management process and, in fact, reflect the planning phase of that process. We can start to judge the effectiveness of management action by the way in which the budget is formulated and used.

2 / The allocation of resources that is formalized in the budget reflects the nature of the operation, and shows which factors are important.

Both of these characteristics are useful, particularly at the start of a cost improvement program. The first tells us something about the degree to which management is discharging its responsibilities and allows us to draw inferences concerning the likelihood that a new program, such as cost improvement, will be successful. The second is important for the obvious reasons that it helps direct our effort and gives us the perspective we need in the allocation of time and attention to match the nature of the operation.

As a general statement, the management structure of an organization corresponds roughly to the cost center structure of the accountant. The usual practice in budgeting is to start with the individual cost centers—the smallest, most basic unit—and to have the managers responsible for each submit their estimates of cost for the time period ahead. These estimates are based on a forecast of output, usually related to the sales forecast, and are consolidated by organizational level until the budget for an entire department has been built. It is one of the responsibilities of the department manager to see that this is done, and done in accordance with current policy. As the various subordinate supervisors go over their budgets with him, he has an opportunity not only to review their plans for the future but also to introduce his own point of view in terms of departmental objectives. This is a firm foundation for the type of review called for by a cost improvement program. The essential point here is that not only does a budget express a plan for the future, but it also is a firm first step in management control. For when the supervisor and the manager reach agreement on the budget, subject, of course, to review by higher management, this becomes a commitment and a plan that will be used in the succeeding steps in the management process.

Variable Budgeting

Usual accounting practice is to establish the "budgeted" figures in the various accounts by cost centers, and then to accumulate and enter the "actual" expenditures in the same format. Differences between these are calculated and posted in *variance* accounts. The supervisor and his manager are thus kept informed of those instances in which the financial part of operations are not proceeding according to agreed-upon plans. At this point, discussions will occur as to the cause and cure of the variations. This is a straightforward form of management by exception, and is quite useful. It all depends, however, upon having a well-thought-out budgeting procedure and a good estimating system.

An interesting aspect of the procedure described above is that, when the variance account shows overexpenditure, the responsible supervisor frequently makes the case that the scope of work at first envisioned has changed, or that sales were higher than expected. In other words, the supervisor himself uses the input-output relationship in justification of cost. This is precisely the relationship we want to establish as central for him, so we should emphasize the point. Many times there will be not one budget for one sales forecast, but rather a series of budgets made up for different levels of sales, or some other measure of workload. This is called a *variable budgeting procedure*, and again it recognizes the association between input and output.

Use of Pareto's Law

The initial input analysis should be done using Pareto's law. This has been discussed, but should be elaborated upon. First, the total budget for the department should be considered. This includes the materials used and all the selling and other overhead expense charged to the department. There is no intent here to enter into an extended discussion of accounting practice, but every person who has supervisory responsibility for a cost center should know how his center is treated in the transfer of material and how his share of overhead absorption is determined. It may seem that this is of only academic interest, since the head of a manufacturing area may seem to be bound by the materials specified by the designer or may have little influence on the allocation of activity of many of the staff functions that are charged into overhead. Yet here is where there exist many opportunities for cost improvement. Furthermore, in the absence of a cost improvement program there may exist a sort of fatalistic attitude on the part of the supervisor in manufacturing, because he may regard materials and overhead costs as being beyond his capacity to influence, and thus not demanding of his attention. However, when he is presented not only with a clear statement of the impact of these costs on his departmental budget, but also, in the cost improvement program, with a means of working with others to improve these costs, we can expect a more positive attitude on his part.

EFFECTS OF BUDGET ANALYSIS

A small but growing company decided to split its present operations into two parts, one devoted to an inexpensive line of products sold through discount houses directly to the ultimate user, and another more expensive custom line sold through distributors to the building-trade contractor. The first line of consumer products had inexpensive parts, was produced in large numbers from a frozen design, and had quality problems principally in the area of acceptance sampling of purchased parts. Even these were not severe. The services of the industrial engineering department were in demand to introduce methods for production at lower cost, since selling price was quite important to the marketing of these consumer products. The second class, custom products sold through distributors to contractors, was much more expensive. Flexibility of design and small manufacturing quantities characterized the production of these items. Quality problems lay in the uniqueness of product design and the stringent operating requirements when the product was put in service. This also required heavy involvement of the design and development staff. Selling price was less a factor in the marketing of this product line than were operating reliability once installed and the ability to deliver products on fairly short notice.

Prior to the division, both classes of product had been manufactured by the same people under common supervision. Overhead charges, including tooling, supervision, selling expense, and all forms of technical assistance, had been applied on the same basis to all products. A cost improvement program had been in operation for some time and seemed to be producing a satisfactory result. Most savings had come from improvement in the cost of consumer products, however.

When the operations were split, a supervisor was designated for each product line and the former man promoted. The two product lines were now in effect to be two different plants. In this case, the routine budget analysis made as part of the cost improvement program proved to be extremely interesting. As each new supervisor reviewed his new responsibilities and established his new goals, there arose a demand from them for more detailed information about the distribution of effort of the several technical staff functions, the sales and management involvement by product line, and other aspects of the procedures for distributing overhead charges. The company was relatively small, and elaborate accounting procedures were not seen as being justified; but some sensible estimates could be made concerning the contribution of specific staff functions to each of the two product lines. This had the healthy effect of bringing home to the staff the fact that the money spent in their support had to be recovered in product sales. It also provided an opportunity for the line and staff to discuss mutual problems centered around cost trade-offs. Finally, it gave top management the opportunity to review operating relationships in a systematic manner. Perhaps these things would have happened without the cost improvement program and the concomitant budget analysis, but the fact is that they did occur within this framework. The critical notion is that the entire budget should be analyzed, item by item and cost center by cost center.

As an example, considering also Pareto's law, let us look again at the electrical distributor discussed in Chapter 9. There only operating expenses were considered for our discussion. However, when the entire analysis is made, we see the following (for major accounts only):

Dollars	Expense	Percent of Total
$ 323,400	Operating expenses	16.2
200,000	Budgeted for growth, profit, etc.	10.0
1,476,600	Purchased material	73.8
$2,000,000		100.0

Here we see that the money spent for purchased material constitutes almost three quarters of the total budget. This means that inventory management should be the first concern of the management. Their effort should be concentrated here, in working toward matching the inventory to customer needs in a manner that will minimize carrying charges and purchase price. It is interesting that monthly carrying charges for inventory amount to about 2 percent of the dollar value, as a rule of thumb.

Another example may be helpful. A major airline undertook a cost improvement program. A Pareto analysis of the budget showed that a substantial item of cost was jet fuel. This did not seem to be an item with much potential for improvement, because the fuel consumption characteristics of the planes were known and measures were in effect to ensure that best practice was being used. It was felt, further, that the price of fuel was essentially an administered price and that no flexibility existed there. However, one rule of cost improvement is that no item should be dismissed without investigation. The matter was discussed in detail by airline and oil company personnel. It was found that certain steps in the treatment (filtering, settling, etc.) of the fuel were being performed both by the airline and the oil company. This duplication of cost and effort was resolved, and substantial cost improvement resulted.

The thing to remember is that, no matter what has been the history of an item of expense, it should not be dismissed without a new investigation. When we accept a cost as "fixed," we decrease our opportunities by that amount. And what set of criteria will we use to write off this opportunity? It seems to be best practice to look at everything.

Another example occurred in a large company in a basic industry. One item in the budget was the cost of the information system, which included several computers and a large number of well-paid people. Partly in self-defense, but principally as a matter of conviction, the manager of this department has developed an agreement which has allowed improvements to the information system which will result in operating economies to be evaluated the same way as any other cost improvement project. Again, an example of considering the entire budget and of giving credit jointly to staff and operating departments when such projects are undertaken.

In summary, we should conduct a thorough analysis of all inputs to our system. Whatever the characteristics of our processor, we know that if we start a system analysis with a hard look at inputs and outputs, we will have unit costs as bench marks, and we will be in a better position to understand the processing function of the system. Input analysis also has the desirable side effect of establishing the cost improvement program within the framework of accounting cost centers and management organizational structures. This gives our cost improvement effort the credibility it must have, since it allows management to treat cost improvement as it does any other management program.

11

Analysis of Outputs: Products, Services, and Work Units

In previous discussion the point has been made that money and other resources, which usually cost money, are used almost universally as the measure of input to a system that produces products or services. It is a fairly obvious statement to say that everybody understands the implications of dollar figures, whether relative or absolute, and thus a Pareto analysis of the budget becomes a meaningful, quantitative statement of the input to such a system. The definition and evaluation of the output of a system may be much more obscure, however. For example, how would one do this for a maintenance department, a police force, or the inspection department of a production shop? Even where we seem to have fairly clear-cut units of output, as is true in some intermediate operations in a production shop, how do we know that the operations being performed are all worthwhile in the sense that some customer will pay us for them? And finally, how can we establish a cost-per-unit "bench mark" in our approach to cost improvement if we do not have the "unit" as well as the "cost"? It is quite apparent that, concomitant to our systematic budget analysis to establish the input to our system, we must also do a systematic and thorough analysis of output. We do this for two compelling reasons:

1 / We must establish relative costs per unit to measure relative improvement.

2 / Consideration of the outputs, from the ultimate product or service we sell on the market to the modifications made in individual cost centers, is usually most productive of ideas for improvement and most useful in involving the entire organization in the cost improvement program.

The most logical way to start the discussion of output analysis might be with a statement of what each supervisor or manager who has cost center responsibilities is asked to do. The specific instruction might be as follows:

List the products, services, or work units of output that your operation produces. In other words, what is it in the way of output that requires for its accomplishment the input of money and resources represented by your budget?

This statement might be followed with the suggestion that the supervisor simply consider the cost accounting concept of matching dollars against units. This may sound straightforward, and in fact in those cases where a well-defined product exists together with a good standard cost system, there should be little difficulty. But as the concept is applied across an entire organization, problems of consistency arise. Unlike dollars of input, for two reasons units of output have little commonality:

1 / Differences in the basic character of the output—for example, varying from well-defined units in production to loosely specified maintenance work to more nebulous outputs such as those in public relations.

2 / Differences in levels of organization—for example, the level of detail of work units of a small section of a department is not and should not be comparable to that of the department as a whole.

SELECTING WORK UNITS

To take these basic differences in turn, the first is probably the more important, and certainly will require the most attention. It also is the most rewarding for the simple reason that in many cases this will be the first searching analysis of these outputs to be done by cost center with a view toward cost improvement. This is particularly true of those work units which are by nature quite variable and unpredictable. Ideally, we would like to set up criteria for the selection of work units that will be similar for all the uses to which we put such measures of output. These uses include all forms of work measurement, short-interval scheduling, cost accounting, estimating, and cost improvement. As a general set of requirements, we should select work units with the following characteristics:

1 | Easy to count.

2 | Consistent with existing information systems.

3 | Covered by historical data, if possible.

4 | Key units, in that other work units can be associated with them in a fairly regular pattern.

5 | Units account for most of the input of money and work for the cost center or group of people involved.

To help us select work units of output, it also is useful to classify them by the concept of variability, which is the point of our previous discussion. The variability that concerns us occurs as differences within what is ostensibly one definition of a unit of output. For example, we have described a maintenance operation in which the maintenance work order (MWO) was used as the basic unit of output. The number of hours of work within these MWO's varied quite widely, however, as discussed in Chapter 10. It is reasonable to suppose that the wide variation in input hours to accomplish an MWO would be the result of a correspondingly wide variation in the character of the work units represented. This in fact turns out to be the case, but the nature of the work units in maintenance usually is not very well defined. Thus, although hours of input may be available, unit costs are almost impossible to develop and comparisons among MWO's are difficult.

For example, two MWO's for the pipefitters' craft were estimated to require the same number of craft hours to accomplish. The work was quite different; one involved a large amount of two-inch steel pipe and the other a much smaller amount of the same pipe. The differences lay in the location of the jobs, the interference with other piping and equipment, and the character of the piping systems (expressed in a code). The influence of these variables may be substantial, and in some cases has been reduced to tables of standard data. But the MWO may not reflect this, and the details that cause time differences are not documented. The experience and judgment of the estimator are important in making the estimate. This is not in itself a "bad" thing. But it should be obvious that it is difficult to develop prediction equations in the absence of documentation of the conditions surrounding the jobs we use for our data base. And we cannot treat the comparisons involved in cost improvement appraisal as if they had the reliability of time standards set on repetitive jobs. This may lead to a tendency to advance the point of view that the work units of output are "too variable to be used for management purposes," and therefore "nothing can be done."

Dealing with Problems in Work-Unit Definition

First, it should be pointed out to supervisors that work in indirect areas is in fact quite variable and to a certain extent unpredictable. Next, let us recognize that because of these two characteristics we cannot expect to be able to define exactly what the work units of output will be in the future, nor even to obtain much explicit information as to what went on in the past. Because of this, the data base in the form of records of work units of output and hours of input is apt to be fragmentary and probably will not contain the information that the supervisor really needs to establish meaningful work units of output.

In the areas of maintenance, materials handling, engineering, and much clerical work, the immediate reaction of the supervisor will probably be that his work units of output simply are too complex and unpredictable to fit exactly the requirements of the cost improvement program. The key word here, however, is "exactly." We should take the position that this is just the first step in analysis, and that there is no preconceived notion of the nature of further analysis. It may be that we will discover relationships precise enough for our purposes in the data that we either have or can get in a sampling of records. We should reassure the supervisor that we are aware of the nature of his work and are accepting this as a fact of life. Our objective is always the level of measurement and control that is practical, and not always the same level.

The fact that this may be the first time that a systematic attempt has been made to analyze the variable operations also is a contributing factor to the supervisor's uneasiness. Traditionally, the staffing of indirect operations has been a matter of negotiation and judgment. We certainly must depend upon the judgment of the supervisor and respect his proven dedication. But at the same time he must realize that we have not extended this staffing procedure into the areas where unit costs are available, even though we presumably have supervisors there who are dedicated and competent.

The first hurdle, therefore, in indirect areas of activity is to get the supervisor to accept the fact that his output can be defined and quantified in a meaningful way. This concept probably is foreign to tradition. And the supervisor would prefer to have a minimum of restriction on his freedom to act. But we can start by pointing out that this is a universal requirement and that management has given it their strong support. Furthermore, the cost improvement program itself requires that he establish specific goals and meet them; the only way he can do this is to compare unit costs over time. Indeed, the initial analysis may well indicate the direction of improvement and should be helpful to him in explaining the program to his people. These arguments may not have as much force at this time as they will later, when both will seem quite sensible to him.

Work Units in Direct Areas

Where work units of output are well defined, work unit analysis has probably been done. In manufacturing, the concepts of standard time and standard cost are well established. Most shops are now scheduled in terms of units of output. The supervisor need only go to the established documents of schedules and standards to get the information he needs. These records will also record scrap loss and various nonstandard units; the supervisor's approach to improvement will be guided more by careful analysis of variance accounts than by a search for causes of variability in the work unit. This is a different problem, but the manufacturing processes are more standardized and some of the improvement probably has been made in the past. The common thread is that we seek improvement everywhere, recognizing that it will take many forms.

DETERMINING WORK UNITS: AN EXAMPLE

An example of a work area in which we start from scratch in the introduction of a cost improvement program might be helpful at this point. This example is typical of many areas in which the initial analysis of work units of output will be sketchy, will probably include items of widely varying scope within some of the individual classifications of work units, and will at the start be considered somewhat artificial. Our example is taken from a large city that was engaged in a cost improvement effort. It was required that each supervisor determine work units of output for his area of responsibility. One such area was that of the Department of Traffic Field Activity. The procedures used may be discussed better if we consider Table 11-1 on page 90.

Notice first the use of the word "budget" in the table title. This is significant in that it expresses the concept of planning performance by output to match the financial planning of input, which is expressed in the dollar or financial budget. This is a valuable concept, for without it we are faced with a series of complex negotiations made on the basis of past practice and present organizational political power. Even worse, if we take only the input or financial budgets in this case, we are ignoring the really important factor of what we are getting for our money. This is of obvious importance in government, where the decisions on budgets are a matter of priorities in allocating resources. No one is against health care or education, for example, but the amount of money devoted to each is a matter of judgment. If we can support these judgments with performance measures, we are bound to make better decisions. The same holds true, incidentally, for indirect and service activities in private industry.

The first line under the "field activity" designation is "borough supervision." This item will match a similar entry in the fiscal budget. No work units are shown. As a general statement, the way in which the supervisory function should be handled is simply to list the time and cost, as is done here. We should not become involved in specific work units, because the supervisor's performance is judged by the success of his organizational unit in meeting goals set for that unit. We can analyze his job to help make him more effective in the use of his time, and we can examine his duties to see that they correspond to the duties given in his position description. But just as the president of a company is judged

Table 11-1.   WORKLOAD AND PERFORMANCE DATA,
BUDGET FOR THE FISCAL YEAR 1960-1961

Program Activity and Subactivity	Work Unit	Workload 1958-1959 Actual
DEPARTMENT OF TRAFFIC		
Fabrication of sign blanks and accessories	Items fabricated:	
	Supports	9,907
	Bracket straps	39,795
	Sign blanks	2,390
Miscellaneous shop work		
Field activity		
Borough supervision		
Installation and maintenance	Signs installed	
	With new supports	14,885
	On existing supports	15,556
	Signs repaired	6,728
	Signs replaced	34,059
Miscellaneous field work		
Accruals		
Program total		
IIIc(3)–markings		
Borough supervision		
Pavement marking–installation and maintenance	New lines painted (ft):	
	Longitudinal	863,264
	Transverse	445,177
	Lines repainted (ft):	
	Longitudinal	9,822,012
	Transverse	1,023,851
Miscellaneous field and shop work		
Accruals		
Program total		

Workload		Performance	Personnel	
1959-1960 Estimated	*1960-1961 Estimated*	*Units per Man Day*	*Man Days*	*Posi- tions*
			450	2
8,000	6,000			
25,000	25,000			
3,000	750			
			1,290	6
			3,655	17
15,000	15,000	4.5	3,333	
25,000	25,000	8.4	2,976	
6,000	6,000	8.	750	
53,000	53,000	7	7,571	
			14,630	63
			1,935	9
				4
				123
			1,935	9
1,200,000	500,000	2,000	250	
800,000	300,000	500	600	
12,300,000	13,350,000	3,000	4,450	
1,200,000	1,500,000	600	2,500	
			7,800	36.2
			1,462	6.8
				2
				54

by the bottom line of the profit and loss statement, the supervisor is judged by the performance of his entire unit, usually measured by the production and cost accounting reports that ordinarily pertain to his area of responsibility. It is his job to direct the activity of others, and we should hold him to that.

To continue, the work units for this particular shop centered around traffic signs. The first type of work unit was "signs installed: with new supports." When the supervisor offered this, he said that it met the criteria for selection of work units (discussed previously), but he did not feel that it was helpful to him in scheduling his day-to-day work. He pointed out that one work unit (sign installed with new supports) might be a small triangular YIELD sign attached to a metal stake, which could be installed in a very few minutes; another work unit bearing the same description (sign installed with new supports) might be a large illuminated display spanning an expressway and requiring several man-days to install. Yet both signs would be counted as one work unit. This is of course true, but two facts should be kept in mind: (1) the product mix, or distribution of size and complexity of signs, probably remains fairly stable from year to year, and (2) our objective here was to establish the initial input-output relationship in terms of gross units for which historical records exist.

It was also stressed that the designation of work units is the supervisor's responsibility. If he wanted to use more precisely defined work units, he was not only free to do so, but was encouraged to do so. One pitfall to be avoided is that of telling a supervisor what his work units are. He knows more about this than the industrial engineer in most cases, and he should get involved himself in any case. The supervisor should use the criteria for selection of work units and do the best he can. In many cases the selection of the work unit is limited by a lack of historical data. This is hard to remedy, but sampling of records may provide enough information to expand our choice. For example, in the situation we are discussing, the supervisor suggested that a simple classification by size would be helpful in refining the work-unit classification. This was done; the sampling revealed that the idea was good, and also that the product mix was stable. Furthermore, the sampling had the valuable side effect of opening up to question certain scheduling and administrative practices, the investigation of which seemed to promise some cost improvement.

The next column in the workload budget gives the reported volume of each work unit for the most recent year or accounting period for which data are available. In our case, this volume was 14,885. Regardless of the degree of detail or accuracy (remember that these records probably were kept for other purposes), this is part of the bench mark. The next column is the supervisor's estimate of this work unit's volume for the next year or accounting period. In our example, this figure is 15,000. When we compare this with the previous volume of 14,885, we can conclude that the supervisor thinks that next year will be much the same as this year. But simply requiring this estimate to be made is important, because it introduces the additional step of considering the workload. This may seem artificial, but in the line below, where the work unit is "signs installed: on existing supports," the value is quite evident. This year's actual volume for this work unit was 15,556, but next year's volume is estimated to be 25,000, a substantial increase. Thus, instead of simply asking for more money, the supervisor has some justification in that the workload has demonstrably increased. This concentration on input-output relationships is consistent with our objectives in the cost improvement program, but it also is sound management practice.

Requiring that this type of workload estimate be made is not new for the production shop; this is how they schedule the shop in any case. But for indirect work and service activity, the use of this approach is new. The basic point is that forcing consideration of output by formal forecasting ties together output with the forecast of input in a way that makes it easy to derive unit cost. This is a more rational approach than the consideration of input alone, because the output requirements determine the input.

One more thought should be expressed here. There are cases in which the work unit itself is almost impossible to reduce to a neat quantitative measure. This is particularly truc in service areas, both in and out of government. What should be the work units for a police patrol force, for example? What measures the output of a quality-control group? In the first case, the number of arrests is easy to count, but perhaps we want our police officers spending more time in preventive work and community relations. In quality control, the quality level in production governs the activity of the people doing the quality control; so if the level of quality in the shop is superb, less inspection usually is required. In such situations it is reasonable to derive less precise measures of output. This may take the form of a multifactor work unit, or it may involve appraisals of opinion, or it may be developed as a relationship to other units of output or other activity in the shop. Service activity may be related to the number of machine setups, the number of new products, or the number of complaints concerning the service. The important concept here is that, although we may not be able to come up with a nice, neat measure of output, we should be able to make some judgments about what creates work for the people involved and to evaluate this.

The next column in the exhibit after the forecasts is "performance," which is expressed in units per man day. This figure is obtained in a straightforward manner in most cases simply by dividing the work units recorded last year by the time recorded to accomplish these units. The difficulty here is that in many cases the record keeping will not have been done in a manner which enables us to determine this easily. However, techniques such as multiple regression and linear programming sometimes enable us to establish input-output relationships from records that are on the surface quite confused. At worst, we will have to consolidate groups of work units, although this will add uncertainty to our values. We also might use work sampling to get information to help us distribute the input in terms of man-hours. This will probably require us to keep more precise records of output during the course of the study, but this is not difficult. But we usually can expect some problems in matching input and output where the work unit is loosely defined. In our example, we should be able to install more than 4.5 small YIELD signs in a day, but we could not reasonably hope to install even one large illuminated sign in the same period. In production shops we have labor standards and payroll and production records, which make it fairly simple to match input and output.

IMPORTANCE OF ESTABLISHING UNIT COSTS

Although the problems discussed are real, the advantages of establishing the unit measure of performance, and establishing it at the very beginning of the program, are also real and very significant. These advantages fall under two general concepts:

1 / The absolute value of the unit of performance in terms of man-hours or dollars is invaluable in making initial judgments concerning the need for further investigation. For example, if our unit cost to prepare an invoice is $50, our intuitive judgment may be that this is "too much," and we will mark this as a subject for further study. Or if we find that the cost per small home appliance repaired is greater than the revenue from this repair, as one utility did, we can abandon that activity. But at least now we know what these unit costs are, and we can make better decisions concerning whether or not the work unit should be analyzed further.

2 / Unit costs are important in the relative sense. We may not know what the unit of output "should" cost. Perhaps it is unique to one of our operations, or perhaps we have determined that it is something that has to be done, regardless. But we do know that the measurement of cost improvement is a relative thing, and we know in which direction lies improvement. So the unit cost is a necessary measure in the evaluation of improvement; as such, it is vital to the cost improvement program. It is one half of the equation for determining cost improvement.

In summary, the analysis of work units of output is a necessary and rewarding step in our procedure for cost improvement. It gets the supervisor involved, because he has the responsibility for defining work units. It is a simple concept, well within the capacity of everyone concerned. It is necessary to the appraisal of results. Finally, it focuses attention on the basic input-output relationship, which is the foundation of the improvement itself.

12

Analysis of Input-Output Results

We have said several times in our discussion of both input and output analysis that each helps define the nature of the operation being considered, and each may give an indication of the avenues to improvement and of further specific actions that might be taken to stimualte changes for the better. At the same time, we have cautioned against forming judgments on the basis of consideration of just half the problem, that is, either input or output alone. Now we have discussed both. In fact, the example in Chapter 11 demonstrated how input and output should be taken together to obtain a cost per unit, which, when arranged by organizational area and by type of input and output, will depict the basic nature of each cost center or group of cost centers. Cost per unit will also provide the bench marks we need to measure relative progress in the cost improvement program.

Our discussion in this chapter will be organized by type of operation: direct labor intensive, indirect or service labor intensive, and material cost intensive. We shall then suggest measures for cost improvement which have proved in the past to be successful when applied to these types of operation. We know of course that most organizations are mixtures of all three, and that no neat dividing lines make sense. We must, however, have some framework for our discussion, and are confident that the reader is capable of relating the discussion to his own situation.

Types of Production

First let us consider the case of a production shop. We know that no two shops are the same, and that the specifics of input and output vary so widely that no generalization is really valid. But to provide a basis for our discussion, we can classify production as being typical of a process industry, fabrication industry, or light manufacturing or assembly industry. These are arbitrary classifications, of course, but may be convenient to use. Examples from the process industries include steel mills, chemical plants, or pulp mills. Examples from fabrication industries include machine shops, furniture factories, and appliance factories. Examples from light manufacturing or assembly industries include electronic assembly plants, clothing factories, and costume jewelry factories. The differences among these types of industries are most easily characterized by examining the three basic inputs of labor, material, and equipment investment.

PROCESS INDUSTRIES

If we consider first the process industries, we almost always find that the heavy charges come from recovering the costs of their capital equipment and the energy and utilities necessary to operate this equipment. Also, process industries are usually involved in some form of mass conversion of materials from one form to another. Process plants usually are large, but direct labor expense, although a considerable amount in absolute dollars, is not predominant; for example, a large department store in a major city may have more employees than a steel mill with vast facilities spread over hundreds of acres. The capital investment in most process industries is immense, and the cost of financing the huge items of equipment is so great that the first emphasis in cost improvement usually is to keep the equipment downtime at a minimum and to operate the facilities for a maximum yield or output. The term "yield" reminds us that we are usually converting material in large volumes from one form to another. We therefore concentrate on obtaining the maximum amount of finished product from each bit of incoming material.

The techniques that have proved to be useful in the detailed-analysis stage range from well-known quality-control techniques, such as the process capability study, to more complex mathematical models, such as optimum-seeking methods and linear programming. In other words, the operator of process equipment is responsible for careful attention to specified procedures in what have become heavily automated items of equipment; but the engineer designs the equipment and instrumentaion, and specifies standard operating practices and the quality and quantity of material, energy, and utilities inputs.

Knowledge of Product, Process, and Customer: An Example

A paper products mill is an example of a process industry. The machinery is massive, the output measured in tons, and the proper adjustment of equipment a vital factor in production cost. A simple yet classic example of the power of exact knowledge of a process came to the author in the course of work in applied statistics done at a manufacturing plant, which used the output of two different paper mills to manufacture punch cards. In the card plant, a Pareto analysis of the input costs showed that the light cardboard stock, purchased from two different vendor mills, was the most significant single item of manufacturing expense. One important characteristic of the card stock was its width. It was purchased in the form of rolls. A sample of about eighty rolls was drawn from the material in the warehouse, which had been accepted by the card plant quality control. The specification for this width (coded for ease of calculation) was that the nominal width was twenty units, with a tolerance of plus four units and minus zero units. The interpretation of this specification is that narrow rolls are really unacceptable, because they create manufacturing difficulties, but that wide rolls are undesirable only because they result in more edge trim waste or scrap. The samples were taken by the customer, in this case, to assist in purchasing the material at the best price and best conformity to specifications. The interesting point is that one vendor in the paper mill, a process industry, turned out to understand his own process much better than the other mill, and used this to his considerable financial advantage.

The results of the sampling are expressed in two histograms, labled Figures 12-1 and 12-2. The characteristic measured, width of roll, is really a measure of the precision of the last operation in the paper mill; this was a slitting operation in which very wide rolls of paper (stock) from the paper machine were slit by a circular knife arrangement to the proper width for sale to the punch card plant. The variability and accuracy of this slitting operation depended upon the state of adjustment and maintenance of a large piece of process equipment common to both vendors. The variability of most mechanical operations such as this is the result of a stable system of chance causes (small eccentricities on rotating parts, looseness in bearings, etc.). The usual form of this distribution of errors is the normal curve, shown in Figure 14-16. [*See Fig. 14-16 on page 168.*] Without giving an extended discussion of quality control, we can say that the measure of dispersion on such distributions is the standard deviation, and that this describes the process capability of the operation. Most commercial tolerances are defined in the United States as including at least 99.73 percent of the items, which is seen to be plus and minus three standard deviations on the normal curve, taken from the average value of all the items. These limits of the average value plus three standard deviations (the upper natural tolerance limit)

FREQUENCY DISTRIBUTION COMPUTATION

Product __CARD STOCK, VENDOR A__

Characteristic __WIDTH, $\frac{1}{32}$" OVER 4-5"__

Date sampled __7/14__

Computed by __W.J. RICHARDSON__

Date __7/16__

Average

$$\bar{X} = AM + C \cdot \frac{\Sigma fd}{N}$$

$$C = 1(\tfrac{1}{32})$$

$$AM = 22$$

$$\frac{\Sigma fd}{N} = \frac{-33}{82} = -.402$$

$$C \cdot \frac{\Sigma fd}{N} = 1(\tfrac{1}{32}) \times -.402 = -.402$$

$$\bar{X} = 22.00 - .402 = 21.6$$
$$(21.6 \; 32^{nds} \; OVER \; 45")$$

Results

$\bar{X} = 21.6$
(32^{nds} over 45")
$\sigma = 1.25 \; (32^{nds})$

Standard Deviation

$$\sigma = C \sqrt{A - B} = C \sqrt{\frac{\Sigma f(d)^2}{N} - \left(\frac{\Sigma fd}{N}\right)^2}$$

$$A = \frac{\Sigma f(d)^2}{N} = \frac{141}{82} = 1.72$$

$$B = \left(\frac{\Sigma fd}{N}\right)^2 = (-.402)^2 = .16$$

$$A - B = 1.72 - .16 = 1.56$$

$$\sqrt{A - B} = 1.25$$

$$\sigma = 1.25 \times 1(\tfrac{1}{32})"$$

$$\sigma = 1.25$$

Value	Number Found at Each Value	f	d	f(d)	f(d)²
			14		
			13		
			12		
			11		
			10		
			9		
			8		
			7		
			6		
			5		
24		10	4		
23		8	3		
22		19	2	20	40
21		29	1	8	8
20		16	0	0	0
			-1	-29	29
			-2	-32	64
			-3		
			-4		
			-5		
			-6		
			-7		
			-8		
			-9		
			-10		
			-11		
			-12		
			-13		
			-14		

N = 82

$\Sigma += 28$
$\Sigma -= 61$
$\Sigma fd = -33$
$\Sigma f(d)^2 = 141$

Symbols

$\bar{X}$ = True Mean
AM = Assumed Mean
Σ = 'Total' or 'Sum of'
σ = Standard Deviation
N = Total Observations
f = Frequency
C = Class Interval – In Moroney, This is Referred to as 'C'.
d = Deviation From AM – In Moroney, This is Referred to as 't'.

Interval	Limits		% of Data
	From	To	Included
$\bar{X} \pm \sigma$			
$\bar{X} \pm 2\sigma$			
$\bar{X} \pm 3\sigma$			
$\bar{X} \pm 4\sigma$			

Figure 12-1.

FREQUENCY DISTRIBUTION COMPUTATION

Product __CARD STOCK, VENDOR B__
Characteristic __WIDTH, $\frac{1}{32}$" OVER 45"__
Date Sampled __7/14__
Computed by __W.J. RICHARDSON__
Average _____ Date __7/16__

$\bar{X} = AM + C\Sigma\ f(d)$

Results
$\bar{X} = 23.0$
$\sigma = 1.0$

$C = 1\left(\frac{1}{32}\right)$"

$AM = 23$

$\frac{\Sigma fd}{N} = \frac{-2}{88} = -.023$

$C\frac{\Sigma fd}{N} = 1\left(\frac{1}{32}\right) \times .023 = -.023$

$\bar{X} = 23.00 - .023 = 22.98$

USE $\frac{23}{32}$" OVER 45" FOR EXAMPLE

Standard Deviation

$\sigma = C\sqrt{A - B} = C\sqrt{\dfrac{\Sigma f(d)^2}{N} - \left(\dfrac{\Sigma fd}{N}\right)^2}$

$A = \dfrac{\Sigma f(d)^2}{N} = \dfrac{98}{88} = 1.11$

$B = \left(\dfrac{\Sigma fd}{N}\right)^2 = (-.023)^2 = .0052$

$A - B = 1.11 - .0052 = 1.105$

$\sqrt{A - B} = 1.05 =$

$\sigma = 1.05 \times 1\left(\frac{1}{32}\right)$

$\sigma = 1.05 \ (32nd")$

USE 1 FOR EXAMPLE

Value	f	d	fd	f(d)²
26		14		
		13		
		12		
		11		
		10		
		9		
		8		
		7		
		6		
		5		
		4		
	6	3	12	24
25	22	2	22	22
24	31	1	0	0
23	23	0	-23	23
22	5	-1	-10	20
21	1	-2	-3	9
20		-3		
		-4		
		-5		
		-6		
		-7		
		-8		
		-9		
		-10		
		-11		
		-12		
		-13		
		-14		

N → 88 $\Sigma + \frac{34}{36}$ $\Sigma - \frac{36}{36}$ $\Sigma\ fd = -2$ $\Sigma\ f(d)^2 = 98$

Number Found at Each Value

Symbols

$\bar{X}$ = True Mean
AM = Assumed Mean
Σ = 'Total' or 'Sum of'
σ = Standard Deviation
N = Total Observations
f = Frequency
C = Class Interval – In Moroney, This is referred to as 'C'.
d = Deviation From AM – In Moroney, This is Referred to as 't'.

Interval	Limits From	To	% of Data Included
$\bar{X} \pm \sigma$			
$\bar{X} \pm 2\sigma$			
$\bar{X} \pm 3\sigma$			
$\bar{X} \pm 4\sigma$			

Figure 12-2.

and the average value minus three standard deviations (the lower natural tolerance limit) are used as the basis for decisions about the process described, which
concern whether or not the process will meet specifications, and in simply stating what the process can do.

From our discussion and Figure 12-2, it can be seen first that vendor B's
distribution of widths followed the normal curve almost in textbook fashion.
Bearing in mind the fact that all rolls measured had been accepted by the customer, it seems that the distribution is complete and symmetrical. The average
width of all rolls is 23, and the standard deviation is about 1.0. It seems that this
reinforces the definition of natural tolerance limits. It also seems apparent that
vendor B is able to ship the entire output from his slitting machines directly to
the customer. It is true that a few rolls are wider than the specification allows;
but it is also true that the customer's quality control accepted these rolls,
because the wider rolls could be used and were in fact easier to use. Production
people at the customer's plant made it known to quality control that they were
willing to absorb the extra waste (edge trim) in the interest of greater ease in
running and reliability of supply. From the point of view of vendor B, acceptance of the wider rolls had significant advantages, but the key was his ability
to keep his process close to the specification, as measured by the standard deviation, and the perception on vendor B's part that a slight variation in the direction of wider rolls would be tolerated by the customer. Knowledge of his own
process and maintaining his equipment properly gave vendor B the following
advantages:

1 / Vendor B had no measuring and sorting operation to be done after the
rolls had been slit. He simply wrapped and shipped the product.
2 / Every roll could be shipped. There was no scrap problem as long as
the present process capability was maintained.
3 / There was no loss of machine capacity through rework or having to
produce extra rolls to offset scrap loss.
4 / Finally, paper products usually are sold by weight. He was shipping
a wider roll, which made manufacturing easier for him, and his customer was paying him for it.

If we now turn our attention to vendor A (Figure 12-1), we see from the
distribution of his widths that there is a concentration of rolls in the center of
the distribution; if we assume that the original distribution as it came from the
slitter in vendor A's plant was normal, we can see that the tails of the curve have
been cut off or truncated. This means that the rolls each had to be measured
after slitting, and those within the limits—from 20 to 24—sorted out and
shipped. If we make the reasonable assumption that the output from the slitter
was normally distributed, and we know that the mean (average) of the distribution of widths was 21.6 and the standard deviation 1.25, we can get some idea of
the area of the normal curve that was eliminated (suggested by the dotted lines)

and, from that, the amount of scrap or rejected rolls. This would amount to about 10 percent of the rolls being rejected as too narrow and about 3 percent as too wide. From the customer's point of view, this made no difference, but for the vendor, the situation led to the following extra costs:

1 / The cost of measuring each roll, sorting out the scrap rolls, and storing them temporarily.

2 / The loss in slitting machine capacity; almost 13 percent of the output of this machine could not be shipped to the customer.

3 / The loss from salvage and rework, which was indeterminate, but was almost certain to be a considerable cost.

The key difference in this example lay in vendor B's knowledge of both his own process and of his customer's real requirements. Vendor B kept his slitter in better adjustment; the variability of his process as measured by the standard deviation was 1.0 versus 1.25 for vendor A. The point of this example for process industries is that vendor A not only had the problem of poor process capability but also received less money for the product he did ship, because paper is sold by the pound and he shipped narrower, lighter rolls. The root of his problem was his own failure to insist that his process equipment be maintained and adjusted properly; both vendors used the same type slitter.

It should be added that this analysis was valuable to the customer, also. Because for all his internal problems, vendor A still shipped less expensive rolls. So when it came time to negotiate a new contract with vendor B, the customer was able to get more realistic terms in the sense that vendor B's practice of shipping the wider, heavier rolls was discussed and an adjustment made.

Improving Yield

In general, in process industries the engineers and supervisors who are responsible for the technical work in operating equipment will almost always be given the improvement of yield as one of their goals. Not only does this reflect in quality and material usage improvement, but also in increased through-put of the equipment. The first step after the initial input-output analysis may well be the application of a series of statistical and mathematical techniques to establish current process capability and perhaps to improve the output by scheduling in a more productive sequence or varying process conditions in accordance with some sort of optimization routine. Even though the technical people are paying some attention to these things now, the cost improvement program will provide a focus for joint effort with production supervision, which may bring improvement.

Many other examples might be given of improvements to cost in process industries, but the approach suggested is a sensible start. The contributions of the industrial engineer and the operations research analyst are particularly valuable because this may be their first opportunity to work constructively with the chemist, the metallurgist, or the electrical engineer. It is unfortunate but true that many of those whose experience and education are in the physical sciences do not realize that the application of statistical and mathematical techniques by industrial engineers and operations research people can be most helpful in improving process-plant operations. In fact, it is sometimes a blind spot of manufacturing managers that they consider process problems uniquely the province of those whose field is in the specific technology with which they deal.

For example, in one pharmaceutical manufacturing plant, whenever a process problem arose, the production manager called in only the chemist to solve it. It is true that the chemist knows more about the formulations and the chemistry of the process than does the industrial engineer. But he does not know as much about scheduling, process capability studies, and other investigative techniques. So the production manager should consider including an industrial engineer or operations research man on his team. In one process plant in the pharmaceutical industry, for example, an industrial engineer used mathematical programming techniques to set up sequences of products and schedules which allowed the number of batches that could be run in a month on major equipment to be increased from fifteen to eighteen. This was done with no additional expense in equipment or labor and without changing the process, but simply by optimizing the order in which batches were run.

A good starting point for the sort of activity suggested is a review of the stated capacities of process equipment. Most firms in the process industries have standard methods of calculating the nominal capacity of each piece of major equipment. This is not as straightforward as it may sound, however, since many specific assumptions must be made in developing these capacity figures. The product mix, lot sizes, incidence of breakdown and preventive maintenance, availability of material, and the capacities of auxiliary equipment all enter into the final determination. We should review the assumptions at the start to see if they still are valid. We may get some ideas for improvement in doing this. In any event, we should compare the stated capacities with our actual experience. Consideration of both these figures is necessary to establish the bench marks against which we will measure improvement. This is a necessary step in any cost improvement program.

The courses of action suggested by no means exhaust the avenues for improvement for process industries, of course. The relationship between maintenance and equipment availability is obvious. And the effectiveness of maintenance itself should be appraised, not only because the cost of maintenance work is substantial, but also because of its impact on equipment utilization and yield. Maintenance will be discussed later. There are other areas of great promise in cost improvement, such as the handling of materials, packaging, methods of financing equipment, and direct-labor cost.

The cost improvement program also brings with it the opportunity for education throughout the organization concerning the need to utilize expensive equipment in the best possible manner to secure an adequate return on invested capital so that the company can stay in business, make a profit, and be attractive enough to investors to maintain adequate capital formation for new facilities.

FABRICATION INDUSTRIES

Let us turn now to that classification of industry which we have termed "fabrication industry." The examples given as typical were machine shops, furniture factories, and appliance factories. In such operations, the Pareto analysis usually reveals that direct labor, materials, and factory overhead constitute the main items of expense; the actual proportions vary with the industry. Regardless of the actual proportions, these three items are common to all plants and significant. Traditionally, we have associated direct labor with this type of manufacturing, but over the past several years its relative importance has diminished. For purposes of our discussion, however, let us first examine the direct-labor input.

Labor Standards

In most plants of the fabrication industry, we have some sort of direct labor standards. The term "some sort of" is used because these standards may range from those set by stopwatch or predetermined motion time systems, after careful study and methods analysis, to a simple estimate by a foreman or by the application of a ratio of labor versus material costs based on historical data. We should not generalize on the level of precision required of a set of time standards. In some cases the foreman's estimate is the most sensible solution; in others we should do the work necessary to obtain good engineered standards. The degree to which the work is repetitive is really the governing factor here. But we can generalize by saying that time standards for direct labor are essential to good shop management, and that we usually find these standards in the shop.

Time standards not only give us the unit labor costs that we need, but are fundamental to the management process. The sequence of planning, scheduling, execution, and follow-up almost demands that we have standards. We simply cannot do any of these separate activities well where direct labor is involved without some dimensioning of the input of work. As we have said, the more repetitive the work, the more apt we are to find well-engineered standards; but even the most informal estimate of times, used as standards for small machine shops, is necessary for the management of such shops. Thus the first thing to do in the input-output analysis of production operations is to review the structure and assumptions of the labor standards and to examine past performance against these standards. This review will not only give us an idea of the accuracy of cost estimates based on standards for use in the cost improvement program; it will also show if previous history supports the notion that cost improvement through methods and equipment improvement has met expectations.

This last statement may raise some eyebrows, but the fact is that unless labor standards and standard cost systems are used as intended in reflecting change, the projected improvements may not be realized. There are too many well-documented cases in which a methods improvement was made that should have resulted in labor cost improvement, but did not, for us to blithely assume that standards and cost accounting systems work as intended. We should first go back, examine a few changes, and see what actually happened. Those who do this and find no problem may consider themselves fortunate, and simply rely on the "audited" feature of cost improvement reporting, which has been discussed. It is true that most of the time when good standards exist and the foremen and operators accept the cost accounting system we can realize improvements. But there are enough exceptions so that a quick review of the matter is suggested. By "method" is meant the motion pattern and equipment manipulation that is operator-dependent. Equipment-dependent jobs will be discussed later.

Employee Motivation

In those cases where expected cost improvements have not seemed to be realized in the past, there usually exists a problem in the motivation of the foremen and the work force. This has been discussed previously to some extent, but the author has no specific set of techniques to suggest to improve employee motivation. This is so much a matter of a specific work situation that about the only safe generalization is that no specific technique has been brought forward which has much universality. The basic problem is that individuals are different, both in their intellectual capacity and outlook on life. The motivation of any individual depends on so many different factors, such as pay, job security, identification of personal goals and company goals, supervisory attitudes, the economic situation in his family, his perception of the company's interest in him as an individual, the nature of his job, the group dynamics of the people with whom he works, and many other influences on his life both on and off the job.

Over the years, starting with the Hawthorne experiments, many techniques or programs have been instituted in an attempt to find a systematic way to improve worker motivation. Individual incentive payment, group profitsharing, and planned job enlargement have been used to bring more of a sense of accomplishment to the individual. Many programs are designed to make supervisory practice more humane and perceptive. Almost all these approaches have been applied successfully in some specific organization. Unfortunately, these same approaches also have histories of unsuccessful application in other organizations where similar problems seemed to exist. This is not to say that the techniques themselves were at fault, but it does suggest that there are variables in motivation that we do not understand completely, and that this fact can lead to disappointment. It is safe to say, however, that a conscientious and serious effort should be made to consider the personal goals, sensibilities, and self-esteem of the individual employee, and to relate these to the objectives of the institution that employs him. If this is done, by whatever means, acceptance of cost improvement is much more apt to occur. One thing is clear from all experience: the employee bitterly resents any effort on the organization's part that he perceives as an attempt to manipulate his attitudes. We should be honest and open, and make it to the employee's advantage to be cooperative in our cost improvement effort.

Assuming that the matters of application of standards for labor cost control and employee motivation have been considered, we should look at two areas of improvement relating to direct labor: (1) improvement in cost per unit through the introduction of better methods, and (2) improvement in other expenses, such as material cost, equipment utilization, and overhead, which should be concomitant to improved labor efficiency. These are interrelated with the use of standards as part of the shop management process.

Methods Improvement

The first area of improvement, that of applying well-understood concepts of methods study, is important for three basic reasons:

1 / It is visible and brings the reality of the cost improvement program directly to the foreman and the wage-roll employees.
2 / Improvement can be instituted relatively, because no elaborate clearances usually are involved. Only the operator, the foreman, and the industrial engineer are involved.
3 / It has a high likelihood of success; some improvement can almost always be made in methods work. This is particularly true when working with jobs that have not been studied recently.

There is no need here to describe in any detail the techniques of methods improvement. Several excellent texts are available; of even greater importance, most companies have had a lot of experience in this work. Methods work has also been the bread-and-butter of the industrial engineer for years. One proof of this is the steady decrease in the proportion of product cost attributable to direct labor. On an international scale, many labor-intensive products are now produced outside the heavily industrialized countries. So the best advice is for the manufacturing manager to place joint responsibility for methods improvement on his supervisor and his industrial engineer. With such management support, methods improvement effort seems to be revitalizing and can be most rewarding.

Expense Improvement in Other Areas

The second area, improvement in expenses other than direct labor, involves the use of direct-labor standards and the way in which direct-labor standards relate to other aspects of shop operations. This improvement usually results from an investigation of the complete labor standard and a consideration of its parts. The underlying power of this approach is that everything an operator has to do to accomplish his work unit of output should appear in the labor standard. This applies to all activity, desirable and undesirable. For example, if an operator has to spend time in miscellaneous delays, materials handling, or in inspection procedures, this time has to be included in the standard. If extra time is required that was not envisioned when the standard was set, this will show up in the labor variance account, and will become a matter for managment attention. Material usage is also specified by the standard or related documents, and control of material is important to cost.

Complete discussion of each of these is beyond the scope of this text, but it has been the author's experience that given an accurate and comprehensive analysis of such costs and the motivation to act, the typical shop management group is perfectly capable of working effectively toward improving the situation. As a short example, the outputs of two groups of assemblers in a metal fabrication shop were found to be well below standard. This variance was investigated, and it was found that parts needed were not available on time. These parts shortages stopped both lines. An ironclad rule was instituted that no production run could be scheduled until all parts were on hand, inspected, and reserved for the lines. This seems to be only common sense, yet the condition had existed for a long time and was only corrected because the direct-labor variance caught management's attention.

One other aspect of labor standards in cost improvement is the impact of their use on quality of product. First, a good standard will specify explicitly the preferred method, the use of which should result in the production of good product. If the scrap is caused by improper work methods, a basis exists for correction. Second, the output of most production operations is measured in good product; if the material or equipment is deficient, this will be apparent, and there will be pressure from the operator and his foreman to correct whatever deficiencies exist. This is of obvious importance, because it constitutes early warning of a problem and emphasizes the need for action to remedy the situation before succeeding operations increase our investment in defective material.

Finally, the intelligent use of labor standards can uncover deficiencies in scheduling practice. This condition will be recognized most clearly when there is a setup or preparation time necessary to an operation. In this case, there is usually a standard economic manufacturing quantity that should be run. If this quantity is not run, a cost variance will occur. This variance may be plus or minus, but in any case it is undesirable. In many cases, schedules are broken to "run a few parts for a good customer." Faced with pressure from sales, the scheduler may do this; but we should recognize that this is expensive, and use the standards to assess the cost. Schedule changes are too common in manufacturing, but will continue to be so until the cost of such production interruptions is known and charged against whatever benefits are gained from satisfying the one customer concerned. In cases of this sort, there is usually little mention of extra cost or of the dissatisfaction of other customers who may receive late delivery. Unfortunately, the manufacturing step comes last, and manufacturing bears the brunt of extravagant delivery problems, material delays, and tedious design problems. We should at least know what the resulting confusion costs.

In summary, fabrication operations usually have a substantial part of their cost in direct labor, although this is seldom the major cost. We should seek improvement not only in the labor cost itself, but also because direct-labor performance against standards is a good indicator of other aspects of shop management. Further analysis of these other aspects, such as material, quality, and scheduling practice may result in worthwhile improvements.

LIGHT MANUFACTURING OR ASSEMBLY INDUSTRIES

The last type of manufacturing we have termed "light manufacturing or assembly." Examples of this type of manufacturing included electronic assembly plants, clothing factories, and costume jewelry factories. The characteristics of these industries are as follows:

1 / Capital investment in manufacturing equipment is relatively modest.
2 / Inventory control and materials management are quite important to both cost and efficiency of the plant.
3 / Considerable labor is involved.

A Pareto analysis of typical light manufacturing will probably indicate that there is a considerable proportion of expense in materials and purchased parts. This is true because characteristically the light manufacturing plant is not really large, does not have the heavy equipment necessary to perform basic operations, and manufactures products that are close to the form in which they will be sold to the ultimate consumer. For example, an electronic assembly plant might use transistors as components in an amplifying system, but probably would not manufacture its own transistors because of the extensive investment in process equipment necessary to make the devices and the high sales volume necessary to support this investment. Instead, the company probably would purchase the transistors, as well as most of the other components of its product. This mode of operation puts considerable stress on the effectiveness of materials management and production-control personnel. One cannot generalize too much, but light manufacturing and assembly plants usually are characterized by a certain flexibility and by the capacity to introduce new designs and new products to reflect changes in customer taste. The clothing industry is an example.

Cost Improvement in Materials Management

Whether because of the relatively high proportion of cost that is related to purchased material, or because the ability to introduce product change depends so much upon the ability to procure and adapt purchased material, opportunities for cost improvement in materials management are usually substantial in light manufacturing and assembly industries. The fundamentals of inventory management are well known. There is no point here in listing all the objectives. However, a few basic areas seem to yield substantial cost improvements in this type of manufacture.

The first area is that of the information system and the system for physical control of material in inventory management. A good first step is to examine the following aspects of materials control:

1 / The extent to which inventory adjustments have to be made, and the nature of these adjustments.

2 / The extent to which bills of material and "where-used" files are up to date and consistent with each other.

3 / The audit trail, from the institution of the purchase order to the shipment of finished goods and billing the customer.

4 / The inventory turnover rate.

5 / The integrity of the shop reporting system.

The need for this fundamental information is not limited to any particular type of manufacturing, of course. But in small plants with rapid rate of product changes, which are typical of light manufacturing, deficiencies in the basic record keeping and physical control of material are more likely to occur. When we add to this the introduction of new purchased parts, the lack of time to produce well-documented product and methods information, and the lack of product experience that is sometimes present, the problems of material control sometimes become formidable indeed. But these problems must be licked, first, before we can apply the techniques and models that are useful in inventory management. The best inventory model on the most powerful computer does us very little good if our shop and storeroom people are not meticulous in the basic procedures of record keeping and materials control.

In the process of working toward what one executive referred to as the "civilizing" of material control and shop reporting, it is almost certain that individual horror cases will come to light. For instance, such things as forgetting to record withdrawals from stock, material that has seemed to disappear, material returned to stock but put in the wrong place, and the failure to match part numbers with bills of material numbers are all upsetting. Such deficiencies can be reduced to a tolerable level only by patient and exacting institution of a sense of personal accountability in the shop, storeroom, and offices. Part of the problem is usually lack of real concern on the foreman's part for the maintenance of real control and system discipline. Many foremen regard these as simply "paperwork." Yet they themselves suffer when system breakdown occurs.

Cost improvement takes the form of reduced inventory losses, less scrap and rework, and less delay time due to material problems. It is fashionable sometimes to blame "the computer" when one is installed, but good shop reporting and attention to established procedures are necessary to make any system work, whether a computer is involved or not.

Inventory Control

Once we have satisfied ourselves that we have a viable information and physical control system, and in the process probably obtained some cost improvement, we can then apply our knowledge of inventory control concepts and techniques. Good inventory management in all types of manufacturing is almost always of extreme importance because it is a most reliable source of cost improvement. Taken together, all classes of inventory usually constitute the largest asset over which management has full discretionary power. As is true of direct labor, we have realized this for years and have been making significant improvements in inventory costs. But this is a continuing process and changes occur. Even today, many firms do not take full advantage of what is known. Some do not assess a carrying charge against inventory or use the concept of the economic order quantity (EOQ). Others still have not made the basic classification of items of inventory into the familiar "ABC" inventory system. This last is of course still another application of Pareto's law. And many firms with computers do not take advantage of the useful time-series-analysis software, which makes practical the use of forecasting at all levels.

It is difficult to understand manufacturing firms which do not pay attention to professionally managed inventory control. In the typical manufacturing plant, the cost improvement per dollar of inventory reduction is much greater than the cost improvement in most other areas. Furthermore, if we can maintain a satisfactory level of customer service, reduction of inventory causes no personnel problems and is immediate and visible. In the past, it was often said that the basic decisions in inventory control were made by clerks, to whom the concepts of economic order quantities and economic manufacturing quantities were alien. This is no longer as true as it once was, but in light manufacturing the pressures to achieve a workable design, develop a process, order material, and get out production are sometimes so severe that orderly inventory management suffers.

Once the information system audit has been made and the manager gains some confidence in the reports he receives, he should appraise the effectiveness of the inventory control system. One simple measure of effectiveness is to look at the schedule changes in a given time period. Some of these will be the result of materials supply or quality problems. Another simple yet informative measure is to examine the number and nature of customer back orders. This measure puts in perspective the real business of any company, which is to compete in the marketplace. In addition, a common complaint of production managers when faced with dollar restrictions on inventory is that, if inventory is reduced, customer service will suffer. A very reasonable response to this is to measure the current level of service, as a bench mark of output. As with many expenses, we tend to concentrate on the input side of the cost equation in inventory management. The output side includes measures of customer service and measures of shop operating losses due to material problems. The turnover rate alone is not a really good measure.

We shall close our discussion of this topic by remarking that, regardless of the type of manufacturing, materials management is a key factor in cost improvement. This is true not only because of the substantial amount of money involved, but also because by following material through a factory we are able to see how effectively shop supervision manages the entire conversion from raw material to finished goods. As the material moves through the plant, it picks up all other costs; this provides a good frame of reference for both analysis and measurement. What we are discussing as cost improvement, furthermore, is really the improvement of the effectiveness of the production cycle. Finally, improvement of day-to-day management practice is probably our most significant goal, because most costs are tied to this. It is in the areas of better planning, better scheduling, better execution, and better control that management should make its contribution. Inventory in all classes is the common thread which ties together manufacturing practice.

INDIRECT EXPENSES AND ADMINISTRATIVE OVERHEAD

Indirect expenses and administrative overhead exist not only in the form of service and support activity for manufacturing but also as the main function in service industries and service organizations such as government, health care, and education. We shall discuss purely service organizations later, but much of what is said about indirect expense and overhead activity will apply to organizations which have as their main function the providing of services.

Indirect manufacturing expenses are those incurred in direct support of manufacturing activity. The most common areas of indirect manufacturing expense are maintenance, quality control, materials handling, production and materials control as it affects manufacturing, and the activities of the technically trained people who are responsible for equipment adjustment and tooling. This last group sometimes is a part of plant maintenance.

Whatever their nature, indirect expenses may be recognized as such because they are not as a rule measured and assigned directly to a specific work unit of output or product, as are direct labor and material expenses. In other words, the distinction between direct and indirect expenses rests heavily on the accountant's judgment. One practical definition of the term is that "indirect expense is whatever the controller says is indirect expense." There is truth in this witticism, but we should recognize that the controller really depends upon the industrial engineer in order to classify costs as direct or indirect. If the industrial engineer is able to measure work and assign its cost to a product, the accountant will be happy to classify it as "direct." If the industrial engineer cannot do this, the expense is thrown into the "indirect" accounts; these are then redistributed proportionately as a function of direct labor dollars or hours, equipment hours, floor space, or on some other basis. Again, the problem occurs when the total of indirect expense becomes large, and when it is realized that very little control exists over this expense. The initial input-output analysis, which establishes some sort of unit cost, perhaps for the first time, is a very useful and powerful means of opening up for discussion on a factual basis the important but difficult question of indirect cost improvement.

Using Work Sampling

It is hard to generalize about any one approach to this cost improvement, because indirect manufacturing expense includes such a diverse group of activities. We can, however, consider plant maintenance as typical, and draw on a previous example. It should be no surprise to the reader that improvement came not from any unusual gimmick, but from the exercise of the basic management process of analysis, setting objectives, planning, scheduling, execution, and follow-up. The initial analysis defined the output of this maintenance operation in terms of maintenance work orders (MWO's). To refresh the reader's memory, a Pareto analysis is repeated from the previous discussion:

Hours per Work Order	Number of Work Orders	% of Total Work Orders	Total Hours Charged	% of Total Hours Charged
24.5 and over	58	4.7 } 15.1%	3,094.5	43.5 } 67.2%
8.5 to 24.0	129	10.4	1,687.0	23.7
4.5 to 8.0	149	12.0	894.0	12.6
2.5 to 4.0	195	15.7	631.0	8.9
1.5 to 2.0	250	20.1	440.5	6.2
0.5 to 1.0	460	37.1	367.5	5.2
	1,241		7,114.5	

The initial input analysis gave the labor hours by craft and by MWO. These data did not, however, tell us anything about the effectiveness of the labor input, nor did they give us any clue to the patterns of work activity or the most likely areas for improvement. A work sampling study was therefore taken by the maintenance foremen of their own craftsmen to obtain a more detailed analysis, which indicated the direction improvement should take. Work sampling will be the subject of Part Three, but it may be said now that a work sampling study is a means of analyzing the activity of people or machines; the end result of a work sampling study is a series of percentages or proportions of the total activity analyzed broken into predefined categories of activity. In our case, the work sampling produced the following results:

Category of Activity of Maintenance Craftsmen	% of Total Observations
Travel	14.9
Planning	2.6
Work	40.1
Clean-up job site	1.5
Wait—tool room	0.2
Wait—other	10.7
Personal and idle	11.7
No contact	18.3
	100.0

Without becoming involved in an extended discussion of work sampling, it is apparent that the craftsmen were not working very effectively. Since no observations were taken during lunch, scheduled personal clean-up, or scheduled breaks, the percentages reflect activity during what may be called "working time." Briefly, the pattern of activity was a reflection of a tendency to "take it easy" on the part of the craftsmen, but, of even more significance, of a lack of planning, scheduling, and supervision on the part of the maintenance foremen. Since the foremen had taken the observations, they realized the nature of the problem. And since this was the start of a cost improvement program, they felt that they could be quite honest and open. They had been assured that no punitive action would be taken as a result of this first study, and that improvement was the objective, not fault-finding.

Incidentally, the reader should be aware that a pattern such as this is quite typical of work sampling studies taken of maintenance and construction workers before a concentrated effort has been made to improve their work. Such a pattern usually occurs as a consequence of management conviction that maintenance work is so variable that nothing much can be done to standardize or control it. This is partially true, but it also is true that the work can, except for emergencies, be planned and scheduled and that, above all, good supervision applied at the work site can be a powerful force for improvement.

In our case, the first step in improvement was to have the manager of maintenance and his assistant sit down with the foremen who had done the work sampling to discuss the results. It was obvious to all that better planning had to be done, because the "travel," "wait-other," and "no contact" categories totaled 43.9 percent of the available working time. Some of the "travel" was necessary, and some of the "no contact" was undoubtedly due to job require-ments; but most of the observations in these categories, in the opinion of the foremen, reflected poor planning and coordination. The important point here is that this was a judgment of the supervisors responsible for the work, and that the results reflected somewhat unfavorably on them and their manager. So it became a mutual problem.

The solution was, first, to concentrate on the larger jobs and, second, to see that each foreman made a point to spend more time actually supervising the work at the job site. Finally, the lead craft foreman was required to go to the site and inspect the job before it was accepted, planned, and scheduled. At the same time, a goal was set to reduce "travel" by one half, "waiting" by one half, and "no contact" by one third. If this all could be done, about 19 percent of the craftsmen's time might be converted from nonproductive to productive activity. This would result in a much larger proportionate increase in output, because the "work" category accounted for only 40 percent of the observations. An increase in this category from 40 to 50 percent is really an increase of 25 percent in productive effort.

Over a six-month period, the steps outlined were followed, and another work sampling was taken. Although the goals were not met in full, the "work" category percentage rose to 51.3 percent. This measured input. The number of MWO's closed out rose 17 percent from the expected number for that month. This was substantially in excess of the random variability of this figure, so it was felt that real improvement had taken place. An interesting aspect of this example was that the foremen did not set a goal for reduction in the "personal" time of the craftsmen, even though the 8.8 percent in the first study was time in excess of the time allowed in the labor contract covering craft work. The foremen felt that this figure would be reduced by the steps toward improvement which they had agreed upon, and that they should correct their own problems first. The results bore them out, as the "personal" time decreased by about one third.

This last aspect of the improvement bears further comment. It is the opinion of the author, apparently shared by these foremen, that the typical craftsman, if he feels that his job is secure, will be willing to put in a good day's work. He regards the planning, scheduling, and coordination of that work as management's job, however, and not his. If he has to stand idle waiting for another craft to do its job, he will. If the equipment that he is to repair is not made available by production, he may simply wait until it is available. If the actual work to be done differs from that specified on the MWO, he will not regard this as a matter of real urgency. These statements should raise no eye-brows, because it is management's responsibility to do the planning and coordination, and the craftsmen's job to do the work. As a general statement, which also applies to other indirect work, we usually obtain our cost improvement through better management, particularly at the foreman's level. The example given is quite typical of maintenance, which in turn is typical of indirect manufacturing expense activity.

It might be helpful to discuss one other example of indirect manufacturing cost improvement. A shipping department in a manufacturing plant was included in a cost improvement program. The basic measure of output, it was agreed, had to be related to shipments. But should the measure used be the weight of shipments, the number of packages, or the number of shipping orders? Or perhaps some combination of these? The input was principally labor hours, and the workload, by whatever measure, had considerable variability. Records were quite complete, however. A work sampling was made by the supervisor. At the same time, a special effort was made to segregate the records for the time period, and to record in addition to routine records the size and weight of each package. After a series of analyses, including multiple linear regression, had yielded a fairly good predictive equation, the output was defined in terms of arbitrarily defined sizes (small, medium, and large packages). Then the prediction equation was used, together with the results of the work sampling, to establish rough time standards, sometimes called reasonable estimates or RE's. A short-interval scheduling system was instituted to improve the planning of workload, and over a few months the number of people in the shipping department was reduced from fourteen to eight. This was possible because it became evident that enough work did not exist to support fourteen payrolls.

Short-interval scheduling is discussed in Part Four; it is a straightforward technique that should involve the supervisor and which has been used successfully to schedule systematically work that is inherently variable in nature. But again, in the example described in the last paragraph, the approach to cost improvement was based on a concerted effort to define and classify units of output, work sampling to analyze input more precisely, and the use of straightforward scheduling and follow-up methods to consolidate this knowledge into a tighter, more efficient operation. This is nothing startling, but rather a renewed emphasis on the fundamentals of management, which has almost always proved to be successful.

THE SERVICE ECONOMY

The last topic of our discussion of approaches to cost improvement is the application of these to service industries and to public-oriented service institutions. The service industries and public-oriented service institutions are important to us because (1) they employ over half of the work force, and (2) they have not, as a rule, been the object of systematic cost improvement effort.

In discussing the first point, we should realize that the most common job in the United States is not production operator but clerk-typist. In addition, the trend is toward further increases in the number of service jobs, particularly in government, education, and health care. A century and a half ago, agriculture was the major source of jobs. But tremendous increases in agricultural productivity occurred through the use of mechanization, fertilizers, hybrid seeds, and better crop and land management; this reduced dramatically the proportion of jobs within the economy that existed on the farm. The emphasis then shifted to production or manufacturing, and at the turn of the century this became the major area of employment. The same process of productivity improvement took place in production that had taken place in agriculture. The improvement was brought about by the introduction of more advanced processes of manufacture, standardization, improvements 'of scale, methods improvement, and better management practice—all done within an expanding economy.

These increases in productivity in the factory reduced the need for manpower, in the literal sense of the word, and the new area of employment became the service industries and service-oriented institutions. These are now the major source of jobs. It is unfortunately a fact, however, that productivity in the service areas has not yet been subject to the same trend of general improvement that characterized the history of agriculture and manufacturing. There has been some movement in the direction of increased productivity in relatively isolated cases, and the public pressures for health care and education at a realistic cost are beginning to be felt. But, at the same time, it is safe to say that tremendous opportunities for cost improvement exist in the service industries and in public-oriented service institutions.

Service Industries

By "service industries" we mean commercial activities in which the value to the customer comes through the performance of some task of a personal, financial, transportation distribution, or entertainment nature. Banks, insurance companies, the communication and entertainment industries, product distribution and sales organizations, and transportation companies are examples of service industries. Our economy is so diverse that it is not worthwhile to agonize over definitions, but one characteristic of many service industries is that they cannot inventory their product, in the sense that a customer wants to travel, shop, or make a bank deposit as a matter of his convenience. In other words, the seat on an airplane cannot be saved for tomorrow if it is empty today.

Public-Oriented Service Institutions

In the public-oriented service institutions, we should remember that government at all levels is now an enormous cost factor in our economy, and that education and the delivery of health care absorb a larger and larger share of the expenditures of a typical family. The reader need only consider the deductions from his own paycheck, plus the entries in his checkbook, to put in perspective the relative importance of service-related activity to his own economic situation. He should also consider that he probably uses the telephone, drives a car on public roads, owns insurance, eats at a restaurant occasionally, and shops for food and furniture. He also sends his children to school and hopes to keep reasonably healthy in a safe environment.

COST IMPROVEMENT IN SERVICE INDUSTRIES

The problems of obtaining cost improvement in the service industries which are profit-oriented are quite similar to the problems of cost improvement in the areas of indirect manufacturing expense, and they yield to pretty much the same approach toward improvement. Experience in airlines, department stores and distribution centers has shown that the well-understood management sequence of establishing units of output (quality and quantity), planning, scheduling, execution, and follow-up of work will usually result in solid cost improvement. In addition, the application of value analysis and methods improvement techniques seems to be effective. Work sampling has been particularly useful, as well as more detailed measurement techniques where appropriate. However, one aspect of service work is often seen by the customer and by management as a problem. That is the level of courtesy and motivation to be of service exhibited by the employee. Here the economic facts of life seem to work against really personal service. In brief, personal attention is costly.

Examples of the cost squeeze on personal services are not hard to find. The economies of operation that go with the use of larger, faster jet planes bring with them problems of giving passenger service to larger groups in less time. The economies of the self-service approach in retailing bring with them problems of pilferage and the loss of a sales-oriented clerk to bring personal touches to the customer. The service industries are responding to these new problems, of course, and are having some success where customer volumes warrant capital expenditures and the introduction of new systems of all kinds. For example, computer-based space control for the airlines should give the passenger much better service in the making of reservations and ticketing. Point-of-sale recording devices are improving inventory control practices in retailing. But the obstacles to cost improvement are difficult to overcome, since the cost of people is by definition the most significant part of input, and wages continue to rise faster than productivity. At the same time, the purchase of services tends to be somewhat discretionary, and a potential customer may decide simply to stay at home instead of going shopping or taking a trip.

Effects of Computer Use

One substantial factor in the service economy is the introduction of the electronic computer. The impact of the computer on manufacturing management is considerable, of course, but in the service area we find the computer as the "production" facility. We think of banks and insurance companies as "paperwork" operations, yet there are 100,000 insurance agents. So while banks and insurance companies are beneficiaries of the computer's power to process data, the impact of the computer extends far beyond these. The computer is now becoming a major factor in the management of business, far beyond its operational function of record keeping. The installation of information and management systems that involve the computer, and the development of the software to operate these systems has become a major service industry. When the computer was a novelty, we sometimes described a computer expert as a person who had not yet received his computer. Now, however, the suggestion is to put the computer and its applications on the same footing as any other function with respect to the administration of cost improvement programs. This means a joint effort with other departments, evaluation based on relative costs, and an audit of results. Information systems managers should welcome this approach, because it usually sets up a situation in which they must work with the manager of some line function to obtain cost improvement. These remarks also apply to the large number of service companies that provide programming and systems work on a contract basis. If such work is treated as any other cost improvement project, many of the questions that have arisen in the past concerning the actual effectiveness of the services provided could have been avoided.

COST IMPROVEMENT IN PUBLIC-ORIENTED
SERVICE INSTITUTIONS

Government at all levels, health care, and education are the most important public-oriented service institutions. The basic problem in these institutions is difficulty in specifying the character of the output. This does not mean that cost improvement in these areas cannot be obtained. We have already given an example from government in our discussion of the maintenance shop for the fabrication and erection of traffic signs. And there are many other areas in such service institutions in which the units of output are fairly well defined. The clerical areas of government and the housekeeping functions in hospitals are amenable to a fairly direct cost improvement approach. But in some areas, such as providing police protection, we really know so little about measures of output that the best we can hope for is to allocate our resources better to match what we perceive as the need. For example, although it is well known that there are certain hours of the day during which the demand for police action is at a peak, many police departments maintain uniform manning levels throughout the entire day. This is not to say that no cost improvement is possible, but rather that some of the problems arise from lack of knowledge of the nature of the work. Fortunately, much more attention is now being directed toward research in these vital service areas. We should also remember that the lack of knowledge cuts both ways; that is, demands for "more people" are hard to support as well as to resist.

It would be naive in the extreme to suppose that the dimensions of the problems in the public-oriented service institutions are only technical in nature. The real questions may be political. The budget of any government, hospital, or school board has to represent a series of compromises, expressed in the alloca-tion of limited resources. These decisions have to be political in nature if they are government decisions. Should we spend money on the police or on correc-tion institutions? Should we allocate funds to the street department in reference to the department of parks and recreation? The only sure prediction is that, as the cost of government continues to climb, the pressures will grow to operate more efficiently. Many capable public servants see this and are in fact making outstanding contributions. Even with the difficulties of establishing measures of output, the basic approach given in this book has been used successfully in government. Political considerations still are the overriding influence, however, and perhaps that is as it should be.

Health Care and Education

The areas of health care and education are similar to parts of government in that output measures are difficult to establish. Quantitative measures abound, but the question of quality of service is hard to answer. In the area of health care, the hospitals have done fine work in cost improvement of some standard tasks, but the judgment of the physician still dictates much of the practice in hospitals. When we consider the obvious fact that we are dealing with human life, the cost improvement argument is weak. But as the pressures of cost mount, the realization has come that every detail cannot be a life-or-death matter, and priorities are being set. There is no point here in extended discussion, however, because noncritical aspects of hospital work are receiving attention. In other matters, the professional opinion of the physician is respected. He wants what he considers best for his patient and for himself. This is a difficult position against which to argue.

In the field of education, a similar situation exists. Educators themselves cannot agree on measures of quality of output, and perhaps this is healthy. But the largest expense is always instructional cost, and the basic question is always the quality of instruction. The various teachers' unions, as is true of any union, have as their objective the vigorous representation of their constituencies, and share the conviction that any money spent on education is really an investment in the future. This has an element of truth, of course, but the reality is that there are incremental gains in quality, and at some point we cannot afford to do some things that are "wants" but not "needs." As a general statement, teachers' unions resist any attempt at appraisal of individual merit on the grounds that no valid techniques exist to do this. However, as financial pressures increase we will find that value judgments involved in the allocation of funds for education will increase in sharpness. Again, many capable professionals in the field are striving for more objective measures of performance in the classroom. There is still a great deal that we do not know, but the stimulus of increasing cost gives some hope for progress.

It may seem to the reader that the author has given short shrift to the enormous expenditures in the area of institutional services. In comparison to the attention given to those activities in the private sector of the economy, this is true. The author has had experience in government, health care, and education, and has confidence that, where the political and professional climate is favorable, the cost improvement programs and techniques discussed in the preceding chapters can be applied quite successfully. The parallel with industry is that top management must provide the driving force; this they do not always do. The general problem therefore is not in the area of application of technique alone, but in the motivation of the political and professional managers to effect improvement. Whether private or public, there exist in the successful organization capable people who would be successful in managing a cost improvement program without compromising the objectives of the company or institution. In brief, success is a practical objective where the motivation is present.

To summarize, we have presented in Part Two suggestions and examples of how to translate the initial input-output analysis into cost improvement. No single book could deal definitively with all the techniques of improvement. Most of these techniques have been understood and applied for years. The objective instead has been to discuss the advantages of organized cost improvement programs, which will create the management structure and involve the people in an organization in a systematic effort toward improvement. The scope of cost improvement is broad, but the people in an organization usually will find the proper technique for their situation and apply it effectively, if they want to do so. The problem in cost improvement has been its introduction as a legitimate, socially acceptable personal goal of every supervisor as part of a systematic effort by top management to improve the competitive position of the organization. We have described an approach to this problem that has experienced some success in application.

WORK SAMPLING

Work sampling is a technique for measuring the activity of people or machines. There are many other techniques which are useful in performing the same type of measurement. The determining factor in selecting the most appropriate technique for accomplishing this measurement is basically the degree to which the work pattern to be studied is repetitive in nature. For repetitive work, we use stopwatch time study or predetermined motion-time systems. Both these techniques involve a detailed method description and fairly precise definitions of the work unit of output. At the other end of the scale, when work units of output are very broadly defined and method description and time-keeping practice are not precise, the most appropriate technique may be a simple index of recorded time per gross work unit of output. The point is that a choice must be made of the most appropriate technique for measuring a given situation. Neither work sampling nor any other specific technique is really appropriate for measuring all types of work. So while the emphasis in this section of the book will be put upon the technique of work sampling and its applications and preferred practice in conducting a work sampling study, it should be realized that work sampling is simply one of a number of useful tools in work measurement.

13

Theory and Definition of Work Sampling

A discussion of work sampling is appropriate for inclusion in this book because it has proved to be a powerful tool for the initial analysis of work situations of an indirect nature, as well as a means of auditing direct-labor performance and equipment utilization as part of cost improvement programs. It translates accounting figures for labor costs into quantitative measures that show the pattern of equipment and personnel utilization. It provides direction for further analysis and gives us an understanding of just how the money we spend for people and equipment is being used. It is, in other words, a broad measurement. This is of extreme importance to management, of course, because in the words of Frederick W. Taylor, "To manage we first must measure." It is also interesting that L. H. C. Tippett, who originated the concept of work sampling in the early 1930s, wrote in his definitive paper on the subject, "A rational attempt at increasing the output of any machine can only be made if the amount of productive capacity lost for each of various causes is known. Indeed, without data of the kind referred to, 'scientific management' is scarcely possible." Taylor was referring to measurement in general, and Tippett to the specific form of work sampling, but the message in both cases is clear: measurement in the appropriate degree of detail is essential to the management process.

DEFINITION

It is always well to begin with definitions. The following definition of work sampling was advanced by the author and Robert E. Heiland in an earlier book on the subject (*):

> A work sampling study consists of a large number of observations taken at random intervals; in taking the observations, the state or condition of the object of study is noted, and this state is classified into predefined categories of activity pertinent to the particular work situation. From the proportions of observations in each category, inferences are drawn concerning the total work activity under study. As an oversimplified example, if a group of maintenance men are observed to be "waiting" in a third of the observations made of their activity, we might draw the inference that better scheduling or supervision, rather than increased crew size, represents the most fruitful area for improvement.

> The underlying theory of work sampling is that the percentage of observations recording a man or machine as idle, working, or in any other condition reflects to a known degree of accuracy the average percentage of time actually spent in that state or condition. If observations are randomly distributed over a sufficiently long period of time, this theory is held to be true, regardless of the nature of the observed activity. Work sampling observations may be likened to a series of photographs taken at random times, with the added advantage that the observer is capable of on-the-spot interpretation and classification of what he sees.

> Work sampling, as the name implies, utilizes the well-established principle of drawing inferences and establishing frames of reference from a random sample of the whole. In this case the "whole" is the total activity of the area, persons, or machines observed during the entire period of time over which observations are made. Work sampling is a practical compromise between the extremes of purely subjective opinion and the "certainty" of continuous observation and detailed study. The advantage of work sampling is that the taking of a few random observations can be done economically, usually as a collateral duty of supervision, while other detailed methods of appraisal are more expensive and may require the full-time services of groups of specialists (*Work Sampling*, p. 2).

Over the years, since this definition first appeared, much work sampling has been done, and a great deal has been learned about the technique and application of work sampling. But the definition itself is still held to be a valid one.

Work Sampling, Robert E. Heiland and Wallace J. Richardson. (New York: McGraw-Hill, 1957.)

STEPS IN CONDUCTING A WORK SAMPLE STUDY

It now seems appropriate to translate the general definition of a work sampling study into a more detailed explanation of what is required to accomplish such a study. Although every study is different, a general pattern of activities is followed in the conduct of a work sampling study. These activities may be listed as follows:

1 / Setting the objective of the study.
2 / Defining the population to be studied. This may include either personnel or equipment, or both.
3 / Selecting the measures of output that reflect the activity of the people or machines who are subject to the work sampling.
4 / Defining the period of time over which the sampling will be done. This definition should specify both short- and long-term periods.
5 / Selection of the people who will do the sampling. Wherever appropriate, the first-line supervisor should be considered for this.
6 / Formulating categories of activity for the sampling. These should be designed for ease of observation and consistency with the objective of the study.
7 / Deciding upon the number of observations to be sought. Here the practical aspects of the sampling situation should be the controlling factor, and the imposing of artificial requirements of reliability should be avoided.
8 / Informing everyone concerned with the study. This most emphatically includes those people who are to be studied.
9 / Developing randomized times for observation.
10 / Developing the necessary forms and procedures.
11 / Conducting a trial study over a two- or three-day period.
12 / The actual taking and recording of observations and the presentation of results.

Each step will, of course, be discussed in greater detail. The obvious final steps of interpretation of results and the institution of management action will also be discussed.

An example should help the reader understand the mechanics of taking the study. Figure 13-1 shows a set of categories of activity typical of a maintenance study. Table 13-1 shows the observation form used in this study. These categories and the observation form were printed on 5- by 8-inch cards for ease of observing and recording. Finally, Table 13-2 shows the results obtained from this study. These exhibits are by no means definitive of the work done in the entire study, but they are indicative of the general pattern that is followed.

1. Travel—Walk or ride, loaded or empty, to or from job site. Push hand truck to or from job. Walk to or from stockroom.
2. Assignment of Work or Planning—Inspect job site or parts to determine what is to be done. Study sketches, prints and specifications, work order, procedures. Receive instructions concerning job from maintenance supv., production supv., or engineer. Fill out time cards.
3. Preparation or Clean-up—Obtain, set-up, or arrange materials, tools, equipment for job. Shut-off or lock-out services. Ready room. Aside or dispose of materials or equipment, etc. Remove safety devices. Disassemble scaffolds. Straighten work area.
4. Work—Handle materials, tools, or equipment at job site. Walk within job site. Cut, weld, gauge, inspect, test, paint, etc. Holding or safety stand-by. Wait while co-worker completes a one-man task (circle no. 4 where this occurs). Machine attention time.
5. Wait for Assignment or Equipment—Wait for instructions or information concerning job from maintenance supv., production supv., or engineer. Await availability of equipment to be serviced.
6. Personal—Lavatory, smokes, personal conversations, idle, visits, etc. Includes any observations of break, lunch, or end-of-day clean-up in excess of specified times.
7. No Contact—No observation of employee during specified tour.

Figure 13-1. Activity Categories: Maintenance Work.

Table 13-1.   DAILY OBSERVATION SHEET: MAINTENANCE WORK SAMPLING

Craftsman	Observation Time									Categories and Totals
										1. Travel
										2. Assign Work or Planning
										3. Preparation or Clean-up
										4. Work
										5. Wait
										6. Personal
										7. No Contact
										Total

Remarks: Date _____

                                            Supervisor _____

Table 13-2.   WORK SAMPLING SUMMARY (% BY SHOP)

Code	Activity	C and M	Elec.	Mach.	Paint	Pipe	P.M.	Night	Total
1.	Travel	19.1	21.3	19.1	15.2	16.4	18.7	19.1	18.6
2.	Assign-Plan	5.6	5.7	8.5	4.2	9.1	6.5	6.3	7.0
3.	Prepare-Clean-up	10.3	10.3	3.6	10.2	6.2	3.4	4.9	6.4
4.	Work	49.0	37.9	39.1	44.5	37.2	32.2	50.0	38.0
④	Work-Wait	1.9	1.8	4.3	1.1	12.6	9.2	1.0	5.8
5.	Wait	3.4	2.0	8.2	4.0	3.7	5.6	2.8	4.7
6.	Personal	9.6	11.2	12.5	13.7	11.8	10.8	15.6	11.8
7.	No Contact	1.1	9.8	4.7	7.1	3.0	13.6	1.0	6.9
	Number of Observations	337	732	903	361	822	919	198	4312

A few additional comments are in order concerning the example given in the previous paragraph. The observations were taken by the first-line supervisors. The results given represent the first two weeks of one month of observations. The unit of output was the maintenance work order, classified by craft and type of work. Briefly, the management action taken of a corrective nature was to institute better planning and scheduling procedures. The supervisors were led to the conclusion that deficiencies existed in these areas because of what they considered to be excessive travel and personal time. Since no sampling was done during scheduled periods of personal time, that appearing on the study was in addition to the contractual rest periods. As a result of this study a new system of planning and control was instituted. Specific goals were set for coming months in terms of greater numbers of maintenance work orders to be closed out and a reduction in the proportion of travel time and personal time in the craftsmen. The first-line supervisors, who had participated in the study, took an active interest in seeing that the new planning and control systems worked. A follow-up study taken after three months' experience, showed real improvement in both the output and the work pattern.

The entire effort in the maintenance area came about as the result of a plantwide cost improvement program. This example, which shows the use of work sampling as a natural extension of the initial cost improvement effort, is typical of the use of work sampling as a basic fact-finding tool. The work sampling study also established a bench mark from which improvement could be measured. Work sampling is an extremely versatile technique in work measurement, but over the years its widest application seems to have been for just these purposes: statement of the basic problem and establishment of a point of departure for management efforts in cost improvement. It is also interesting that the first steps in conducting a work sampling study correspond very closely to the first steps in instituting a cost improvement program. In fact, some of the preliminary activities should actually have been already accomplished; for example, the statement of the objective, the definition of the population, the selection of measures of output, and the involvement of the first-line supervisor are all part of the basic approach in instituting a cost improvement program.

FEATURES OF WORK SAMPLING TECHNIQUES

Work sampling, then, has many attractive features that make it a useful tool in cost improvement programs. It also has many other uses, which will be discussed later. Rather than list "advantages" and "disadvantages," it makes more sense to list the things that work sampling can do, and also the things that it cannot do. The following are positive features of work sampling.

1 / Gives an overall picture of the distribution of activities of people and machines.
2 / Involves the first-line supervisor in the appraisal of his own operation.
3 / Is a relatively simple, easily understood technique.
4 / Is particularly appropriate for the analysis of indirect labor.
5 / Possesses known reliability.
6 / Is relatively inexpensive, and is basically a "do-it-yourself" technique.
7 / Does not commit management to the introduction of incentives.

There are, of course, many things that work sampling will not do. We have already indicated that it is not an appropriate tool for the measurement of highly repetitive work, for example. Among the things not usually associated with a work sampling study, we find the following:

1 / Work sampling does not include the basic information necessary for individual methods improvement. Even the most elementary techniques for the development of time standards include some sort of methods description from which improvements can be made.
2 / Work sampling does not give quick answers. The usual sampling period for an initial study is a calendar month.
3 / We do not usually incorporate rating or leveling in the same observation of activity in a work sampling study. This is not an absolute rule, but is valid as a general statement.
4 / We do not get a direct time per unit value from a work sampling. Again, this is not an absolute rule, but is valid as a general statement.

These comments are consistent with the expressed view that work sampling is simply one of several work measurement techniques. We need not belabor the point that no one technique is universally applicable. In Chapter 14 we shall review briefly the major methods of work measurement, so that the reader will understand the criteria used for the selection of a technique appropriate to his problem.

14

Practice of
Work Sampling

WORK MEASUREMENT TECHNIQUES

Work measurement is the process of establishing the input in terms of time of people and machines necessary to produce a given output in terms of accomplishing a work unit or providing a service. This is a broad definition and is intended to be so. Techniques of work measurement range from gross historical standards, developed from records perhaps not originally intended for this purpose, to precise measurement using a stopwatch or predetermined motion times and based upon explicit definition and control of methods, equipment, and environment. There is, in other words, a scale of measurement techniques. This scale is fundamentally based on the degree to which the work is repetitive, defined, and predictable. We shall discuss this scale starting with the broadest techniques, and relate each technique to the general scheme in terms of input-output characteristics. In general, there will be a decrease in the span of time considered from technique to technique.

Yearly Historical Data

Yearly historical data will be considered as the broadest basis for work measurement. A specific example of these data appears in Table 11-1 (page 90) in the part on short interval scheduling. This example, taken from municipal government, shows the basic nature of the data. The categories of work units tend to be very broad and not very informative as to the characteristics of particular work units. There is usually a tendency to "mix apples and oranges." This is true because these data, if they are kept at all, are usually kept for accounting purposes or as an administrative aid. Input data are usually measured in terms of simple payroll figures, and the performance measure, in this case "units per man-day," is generally not useful for controlling day-to-day activity.

However, such measures do have the advantage that they may be traced historically, and they are based on whatever the existing record-keeping system is. At the very least, they establish the concept that attention must be given to both the output and the input in a work system. Such yearly historical data appear not only in budgets, as in this example, but also in various cost accounting records. These records have the following disadvantages: usually no method is specified, detailed timekeeping usually is lacking, and the definition of work units is too broad to be useful in local control. They do have the advantages of cheapness, they represent existing practice, they fit into cost improvement programs, and they serve as a starting point in all sorts of work measurement. If we were to express our measurement schemes graphically, yearly historical data would appear as a single horizontal bar, as follows:

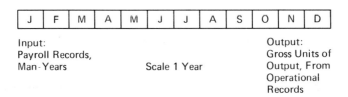

Figure 14-1.

Monthly Historical Data

Monthly historical data are simply a refinement of yearly historical data. The monthly data tend to give more detail, they show whatever trends exist within the year, the relationship between input and output is more easily established because of the shorter time span, and they provide more data points for

Table 14-1.   MAINTENANCE WORK ORDERS

10-Month Summary

1971	Total MWO's	Total Hours Charged	Total "RM" MWO's	Total "RM" Hours Charged	Net MWO's	Net Hours Charged
Jan	1168	5649.0	138	1200.0	1030	4449.0
Feb	1174	6114.0	117	953.0	1057	5161.0
Mar	1295	6964.5	120	977.0	1175	5987.5
Apr	1241	6777.0	122	1019.0	1119	5758.0
May	1164	6725.0	136	1230.5	1028	5494.5
June	1280	6360.0	109	939.5	1171	5420.5
July	1092	5945.0	127	1126.0	965	4819.0
Aug	1243	7114.5	113	907.0	1130	6207.5
Sept	1261	7118.5	146	1377.0	1115	5741.5
Oct	1347	7304.0	125	1012.0	1222	6292.0
10-Month Total	12265	66071.5	1246	10241.0	11014	55330.5
Monthly Average	1227	6607 hrs.	125	1074 hrs.	1102	5533 hrs.

3-Month Detailed

Hours Charged	August % of MWO	August % of Mo. Total	September % of MWO	September % of Mo. Total	October % of MWO	October % of Mo. Total	3-Mo. Avg. % of Total
0.5 to 1.0	462	37.2	440	34.5	442	32.8	34.9
1.5 to 2.0	252	20.3	245	19.4	278	20.6	20.1
2.5 to 4.0	194	15.6	200	15.9	257	19.1	16.9
4.5 to 8.0	148	11.9	165	13.1	174	12.9	12.6
8.5 to 24.0	129	10.4	155	12.3	143	10.6	11.1
24.5 & over	58	4.7	56	4.4	53	3.9	4.3
	1243		1261		1347		

the use of statistical techniques for analysis. Again they have the disadvantage that these records probably were not kept primarily for work measurement purposes. This means that we will have many of the problems associated with yearly historical data in that work units may be grouped together and timekeeping may be inadequate. An example of monthly data has already been given in Table 14-1 (page 135), which shows the monthly pattern of maintenance work orders. The graphic presentation of monthly historical data is as follows:

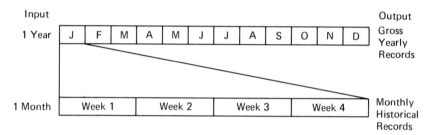

Figure 14-2.

At this point it is appropriate to remark that the raw data of input and output figures usually must be made the subject of statistical analysis in order to produce time standards for given units of work. A very common technique for doing this is the use of multiple linear regression. This technique will be discussed in Chapter 16. It is particularly useful in the typical case of monthly historical data, in which a total time is recorded for a number of different work units, but the detailed recording of time per work unit is not available. Linear programming is another mathematical technique that can be used when the total time and the distribution of work units are known, but an approximation of the time per work unit has to be developed.

A word is in order here about the use of linear programming, multiple linear regression, and other mathematical techniques in the development of standards. This is an opportunity for the industrial engineer who has received an education in these techniques and the use of the computer to apply these powerful tools to a traditional task of work measurement. There are problems here, of course. We still do not have methods description, we still may not have clear-cut distinctions among work units, and there usually remains a large amount of time unexplained by the performance of people accomplishing work units. But the use of mathematical techniques for the reduction of gross monthly data to rough time standards is very definitely in order.

Work Sampling

In the scale of work measurement techniques, work sampling is probably most appropriately placed here, after the monthly historical data. One basic reason for this is that very frequently we conduct our work sampling over a calendar month or an accounting month. This makes it easy for us to match input and output units, because output units traditionally are kept by the month, as are timekeeping records. There is no point in repeating comments already made about the technique of work sampling. However, to improve understanding, a graphical representation of a monthly sample is presented; vertical lines represent the times at which work-sampling observations were made.

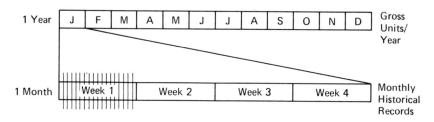

Figure 14-3.

Weekly Historical Records

Many of the same comments that apply to monthly historical records also apply to weekly records. However, the very obvious point should be made that the weekly records, taken over a shorter time span, should reveal more detail. It should be more convenient, for example, to identify particular operations or work units and to get a better grasp of the time required to perform these. From the point of view of the application of statistical and mathematical techniques for analysis, the greater number of data points should be valuable. In the graphic presentation given below, the work pattern is indicated with solid marking for useful work, cross hatching for work that is of an auxiliary nature, and blank space for nonproductive time. The vertical lines again indicate the times at which snap observations might be made in a work sampling study.

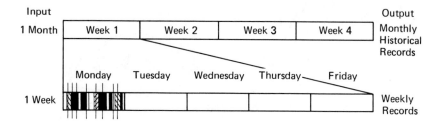

Figure 14-4.

Production Study or Eight-Hour Study

It should be obvious now that all work measurement is a form of sampling. The month is but a sample of the year, the week a sample of the month, and so forth. When we are developing time standards this concept of sampling holds, because regardless of the conditions under which they were developed, we assume that these conditions are but a sample of the entire population, which includes work to be done in the future. We make the assumption that the same methods, work units, and environment will exist in the future as they did when the work first was measured.

This brings us to a technique of work measurement, commonly called a "production" study or "eight-hour" study, in which an observer simply goes to the job site at the beginning of the work day and records everything that happens during the day. This has all the advantages of direct observation, including a direct method description, description of the environment, and timing of the various work units. Two problems exist, however. First, how do we know we have selected a typical day, and, second, how do we know that we have selected a typical operator? We can make these judgments based on our records and our general knowledge. A remaining difficulty is that we know that the presence of the observer in a work situation sometimes biases the situation. Operators will sometimes vary their pattern of work to present the picture that they feel is most favorable to them. The illustration from Frederick W. Taylor's book *Shop Management* (Figure 14-5) is a vivid commentary on this phenomenon.

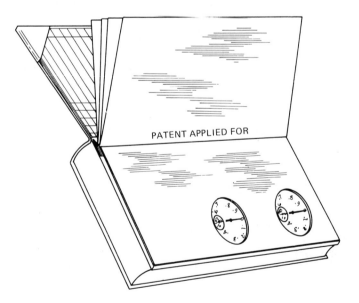

PATENT APPLIED FOR

Figure 14-5.

Of course, today no sensible industrial engineer would ever engage in any form of surreptitious or concealed work measurement. As a general rule, the production study has been replaced by work sampling. The production study is sometimes used in checking a particularly important time standard, however. The graphic presentation (Figure 14-6) shows a typical work day, again indicating the various types of activity and again having vertical lines to show the times of work sampling observations.

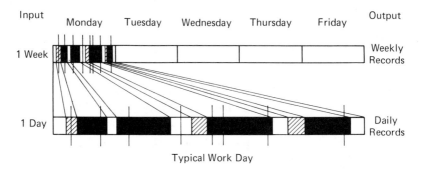

Figure 14-6.

Stopwatch Time Study

Where operators are engaged in essentially repetitive work, the technique of stopwatch time study may be an appropriate one for work measurement. The essential condition for the use of stopwatch time study is that the job itself or substantial elements of the job should be repetitive. Usually, each repetitive job is first studied in some detail and a preferred method arrived at. Then a fairly explicit methods description is made, in which the hand and body motions, tools and equipment, specifications of material input and output, physical qualifications and skill of the operator, and the environment surrounding the job are all described in sufficient detail so that the job may be reproduced. Several cycles of the job then are timed, and some form of performance rating is employed to relate the observed performance to a standard level of performance. Figure 14-7 (a and b) is an example of a methods description and Figure 14-8 of an observation sheet. Stopwatch time study is by far the most common method of measuring repetitive work. Typical cycles range in terms of time from about one minute upward to about one hour. These are not absolute figures, of course, but serve as a guide. When stopwatch time study is done by professionally trained people, it should be quite effective.

Work sampling is used in conjunction with stopwatch time study to establish percentage allowances that are added to the observed and rated times for each cycle. These allowances are designed to add to the work cycle time the

TIME STUDY SHEET

Operation _____

Study File No. _____
Fixture No. _____
Drawing No. _____
Attachments to This Sheet:
TABLE 17, WORKPLACE SKETCH
Date 7/16/74

Part _____
Specif. No. _____
Study by W.J. RICHARDSON
Approved _____

Operator Name _____
Operator No. _____
Dept. PACKAGING
Machine Type and No. _____
Time Began Study 10:22 AM (OBS)
Time Ended Study 10:38 AM (OBS)
                 11:28 PM (DETAIL)

Workplace Symbols	Sketch of Workplace
	SEE ATTACHED SHEET

Elem. No.	Left Hand Description	Right Hand Description	No. of Obs.	Allowed Time
1	GRASP TRAY, TO	WORK AREA Ⓐ, GRASP 2 VAILS (1 PER HAND), TL TO Ⓐ, SNAP IN TRAY, REPEAT (EXCEPT FOR TRAY) 4 MORE TIMES TO FILL TRAY. ASIDE FILLED TRAY TO BELT. RL TRAY	44	.273
2	TE, GRASP SCOOP, FILL LINE TRAY FROM BULK VIALS, (3 OR 4 SCOOPS) (ABOUT 300 IN TRAY)	RL SCOOP	2 (1/20)	.012
3	REACH, 6 A DOZEN OR SO TRAYS AT A TIME, TL TO WORKPLACE, REPEAT, RL TRAYS		2 (1/10)	.004
	"PIECE" IS ONE VIAL.	"UNIT" IS TRAY OF 10		
	.0289 mm, Total Time Allowed Per Piece			.289

Production at Standard 207 TRAYS (UNITS) Per Hour
Production During Study 234 TRAYS (UNITS) Per Hour

Figure 14-7(a).

Sketch of _____

_____

_____

Scale — Each Square = __1'__

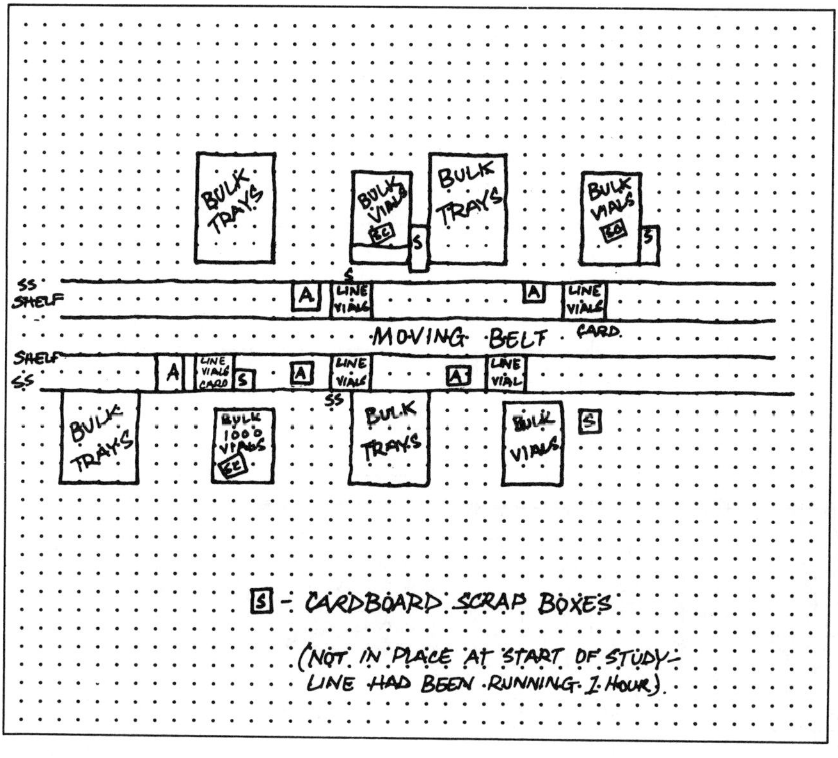

Figure 14-7(b).

amount of time that the employee needs for miscellaneous delay, tool care, material handling, and other factors of the job environment. These are indicated in the graphic presentation as a solid block of time; however, these events occur throughout the day and it is simply convenient to include their time requirements as a percentage allowance. When stopwatch time study is professionally done, an important by-product of the overall time standard is a series of times for the individual elements of the job. These element times later can serve as the basis for standard data. The graphic presentation representing the time span of a typical stopwatch time study is shown in Figure 14-9.

No.	Terminal Point	1 R	1 T	2 R	2 T	3 R	3 T	4 R	4 T	5 R	5 T	6 R	6 T	7 R	7 T	8 R	8 T	9 R	9 T	10 R	10 T	11 R	11 T	12 R	12 T	13 R	13 T	14 R	14 T	15 R	15 T
1	RL TO BELT	00	24	24	25	54	24	78	23	01	25	26	26	52	22	74	25	99	27	26	24	50	24	74	24	98	27	25	26	51	24
2	RL TRAYS	29	34	5																											
3	RL SCOOP																														
4																																
5	RL TO BELT	75	28	18	23	41	22	63	26	89	24	13	24	37	26	63	22	85	23	08	24	32	24	56	22	78	24	02	24	26	22
6	RL TRAYS	04	15																												
7	RL SCOOP	89																														
8																																
9	RL TO BELT	48	28	85	23	03	25	33	30	63	24	87	24	11	23	33	34	57	24	81	24	31	26	67	22	78	25	03	25	28	
10	RL TRAYS	57	9																												
11	RL SCOOP																					88	14 26										
12																																
13																																
14																																
15																																

Symbols Used in Study — **Details of Adjustments**

	1	2	3
	D-S	E-8	E-8
	F-O	F-O	F-O
	H-O	H-O	H-O
	L-7	J-2	J-2
	P-2	O-1	P-2
	W(5)-13	W(9)-20	W(4)-10
	27	31	22

ELEMENT	1	2	3	4	5	6	7	8	9	10	11	12	13	14	15
Total Time in	1074	41	14												
Number of Readings	44	2	2												
Average	.244	.205	.07												
Rading and ADJ	80/27	80/31	80/27												
Base Time	.248	.2148	.071												
1.00 + %Allowance	1.10	.2363	1.10												
Allowed Time	.2728	.2363	.0782												
Rounded	.273	.012	.004												

$\frac{7}{20}$ $\frac{7}{20}$ $\frac{1}{20}$

Figure 14-8.

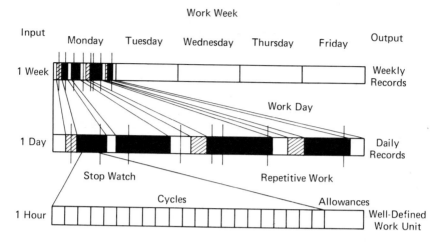

Figure 14-9.

Predetermined Human Motion-Time Systems

When work becomes highly repetitive—that is, when a relatively short job is done hundreds or thousands of times—the sample becomes one cycle. The time span involved is, let us say, about one minute. For such highly repetitive jobs, particularly those in electrical and mechanical assembly, a technique known as predetermined human motion-time systems seems quite appropriate. Such systems either observe or hypothesize a motion pattern, break this total pattern down into discrete movements of the hands or other body members, and assign times for each motion from experimentally developed tables. The underlying theory is that when the conditions surrounding any particular motion can be defined, a time for that motion will be fairly universal. The times for each motion in the described method are simply added up, allowances applied, and a standard derived. Of the several systems of predetermined human motion times, perhaps the most widely used is methods time measurement (MTM). Figure 14-10 is a sample table of values from MTM. These systems have the advantage that a job can be designed and a standard set before the job runs, even when the job is not exactly typical of work previously done in the shop. Also, there is a greater likelihood of an exacting methods review when using predetermined human motion-time systems than when using other techniques. As is true with time study—or with any other technique—the professional competence of the person using the technique largely determines its degree of effectiveness. Since the job already is broken down into elements, the development of standard data, using studies made with predetermined human motion-time systems, is quite straightforward; this is a definite advantage of the technique.

METHODS-TIME MEASUREMENT
APPLICATION DATA

Simplified Data

(All Times on This Simplified Data Table Include 15% Allowance)

HAND AND ARM MOTIONS	BODY, LEG AND EYE MOTIONS

REACH OR MOVE   TMU

	TMU
1" 2	Simple Foot Motion ... 10
2" 4	Foot Motion with Pressure 20
3" to 12" 4 + Length of Motion	Leg Motion 10
over 12" 3 + Length of Motion	
(For TYPE 2 REACHES AND	Side Step Case 1 20
MOVES use Length of Motion Only)	Side Step Case 2 40

POSITION

Fit	Symmetrical	Other
Loose	10	15
Close	20	25
Exact	50	55

	TMU
Turn Body Case 1	20
Turn Body Case 2	45
Eye Time	10

TURN — APPLY PRESSURE

Turn	6
Apply Pressure	20

Bend, Stoop or Kneel on	
One Knee	35
Arise	35

GRASP

Simple	2
Regrasp or Transfer	6
Complex	10

Kneel on Both Knees ...	80
Arise	90
Sit	40
Stand	50
Walk per Pace	17

DISENGAGE

Loose	5
Close	10
Exact	30

1 TMU = .00001 Hour
= .0006 Minute
= .036 Second

Figure 14-10.

It is also appropriate here to state the obvious: many jobs can be measured either with stopwatch time studies or with a predetermined human motion-time system. Those techniques, when professionally done, are effective. It is beyond the scope of this book to make detailed comparisons. The use of a particular technique in a particular work situation is so much a matter of past history, availability of professional help, job environment, and the projected use of the standards that no generalization is sensible. A graphic presentation of the place on the scale of measurement of predetermined human work-time systems is as follows:

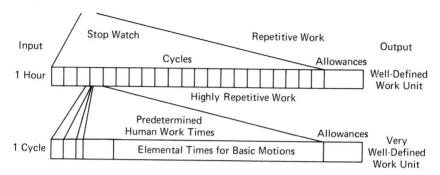

Figure 14-11.

STANDARD DATA

The term "standard data" is applied to time standards generated by taking elemental data from either stopwatch or predetermined human work-time systems, relating these elemental data to the physical characteristics of a task, and assigning a time for that task based on our previous experience. Standard data may be in the form of either tables or formulas, or a combination of these. Basically, if we have studies on the job of drilling four holes in a bolt circle, eight holes in a bolt circle, and twelve holes in the same circle, using the same equipment and general method, we should be able to predict the time of drilling six holes or ten holes without going out and studying the entire job anew. Standard data are used for estimating, setting production standards, developing manning tables, and for the many other uses to which we put work measurement. The development of standard data is a fertile field for the application of mathematical and statistical techniques of analysis. Standard data may not be as precise as individual studies, but they are a better predicting tool, take advantage of pooled opinion, and are extremely economical to use where appropriate. Figure 14-12 is an example of a table of standard data, and Figure 14-13 shows examples of standard data formulas.

A few comments are in order concerning the scale of work measurement just presented. As we go from the top to the bottom of this scale we know more and more about the details of the job. But at the same time we have covered less and less of the entire work situation. For example, we can spread out the weekly time cards of a dozen men and make some sense of what occupied their time during an entire week. This is a comprehensive look at labor utilization, but it does not give us any grasp of methods or much of an indication of how the work could be improved. Generally speaking, in cost improvement programs, we start

HANDLING TIME — COMBINED ELEMENTS DESCRIPTION							
A. PICK UP — PLACE IN POSITION TO HOLD — REMOVE AND LAY ASIDE				WEIGHT RANGE IN POUNDS			
	Less than .5	.5 - 1.	1. - 2.5	2.5 - 5	5. - 15.	15. - 30.	
On table — V blocks or parallels	.0013	.0016	.0021	.0026	.0036	.0045	
In vise or between centers	.0011	.0014	.0019	.0024	.0034	.0038	
In fixture or chuck	.0016	.0020	.0027	.0034	.0048	.0060	
On index plate or rotary table	.0013	.0018	.0024	.0031	.0043	.0055	
In collet	.0013	.0016	.0021	.0026	.0036	.0045	
B. TURN PIECE							
On table — V blocks or parallels	.0005	.0007	.0011	.0015	.0025	.0035	
In vise or between centers	.0007	.0009	.0014	.0018	.0030	.0040	
In fixture — chuck or index plate	.0008	.0011	.0016	.0022	.0033	.0045	

C. GAGE PIECE	DIMENSIONS MEASURED					
	0″–2″	2″–4″	4″–6″	6″–8″	8″–10″	10″–12″
Outside micrometer — plus or minus .001	.0020	.0023	.0025	.0027		
Outside micrometer — plus or minus .0005	.0023	.0025	.0027	.0029		
Depth micrometer — plus or minus .001	.0028	.0030				
Depth micrometer — plus or minus .0005	.0030	.0030				
Scale	.0013	.0013	.0013	.0026	.0026	.0026
Caliper	.0016	.0018	.0021	.0024	.0028	.0032
Woodruff or width gage	.0024	.0027	.0031	.0034	.0040	.0045
Pin gage	.0015	.0016	.0017	.0020	.0023	.0025
Snap gage	.0013	.0014	.0016	.0017	.0020	.0022
Micrometer over 2 pins	.0108					

D. CLEAN OR BRUSH					
Blow off with air hose	.0009	Wipe clamping surface	.0005	Piece — 0 lbs. to 10 lbs.	.0010
Flush piece or fixture	.0009	Wipe button-finger or rag	.0003	Piece — 10 lbs. to 30 lbs.	.0020
Scrape chips aside	.0008				

E. TIGHTEN AND RELEASE HOLDING DEVICES					
Collet or draw chuck	.0005	Tailstock-hob	.0033	Locating screw by hand	.0022
In vise by hand	.0020	Thumbscrew or nut by hand	.0015	Locating screw with wrench —	
Cam lever or toggle clamp	.0006	Locating plug — 1″ under	.0008	removed	.0046
Rack and pinion	.0021	Locating plug — 1″ over	.0016	Holding screw with wrench	.0027
Cover — hinged	.0010	Clamp in place — per clamp	.0021	Nut with wrench — removable	.0051
Toe clam, U-clamp etc — slide	.0008	Chuck — per jaw or universal	.0017	C washer — on and off	.0012
Air clamp — air collet	.0004	Thumbscrew-semi turn to lock	.0004	Allen screw with wrench	.0025

F. MISCELLANEOUS HANDLING ELEMENTS				
Start and stop machine		.0008	Place, remove splash guard	.0013
Engage feed-table, rapid reverse or cross		.0005	Pry part loose from holding device	.0007
Run table–forward or reverse–by hand (Per inch)		.0005	Trip lever or actuate switch	.0005
Run table–forward or reverse–rapid travel (Per inch)		.0002	Back off and advance hob–by hand (Per inch)	.0002
Run table–forward or reverse–hyd. or hand mill			Dog–on and off shaft–tighten–release	.0021
(Per inch)	.0001		Position hob	.0275
Raise and lower cutter head-hand mill		.0010	Reset head	.0075
Raise or lower table–by hand (Per inch)		.0015	Visual inspect	.0002
Run table by hand–per inch — K & T — 2H only		.0010	Place cardboard — per layer	.0020
Apply straight edge (3 times)		.0019	Place paper separator between parts	.0006
Use feeler or shim for locating		.0015	Count	.0002
Unlock, index, lock (dividing head)		.0030	Rub surface flat, under 10″ total 2 dim.	.0012
Set to clip		.0010	Rub surface flat, over 10″ total 2 dim.	.0024
Seat piece or tighten clamp with mallet		.0005	Index fixture	.0009
Each additional tap	.0003			
Brush surface or tool with oil		.0008		

Figure 14-12. From Handbook of Standard Time Data, *Hadden and Genger (New York: Ronald Press, 1964) pp. 142-143.*

146

$$Y = a + (b_1 \times X_1)$$

(predicted time in hours) — (constant term) — (coefficient) — (number of work units of one type)

$$+ (b_2 \times X_2),$$

(coefficient) — (number of work units of the next type)

and so forth to X_6

Actual equation:

Multiple coefficient of correlation 0.49.

$$Y = 53.44 + 0.009X_1 + 0.068X_2 + 0.051X_3 + 0.001X_4 + 0.021X_5 + 0.069X_6$$

t values: 0.69   2.44   1.92   0.17   1.29   1.89

Figure 15-1. Prediction equation based on historical data used (with some adjustments to coefficients b_1 and b_4) in a short interval scheduling application. This prediction equation reflects a loose control situation. (See Appendix.)

with very broad measures to establish the nature and extent of labor input cost. We then go into just enough detail to have the appropriate analysis for planning improvement. So there is no such thing as a "good" or a "bad" technique. We can describe a technique as "good for overall measurement" or "good for detailed analysis." Work sampling is simply one of these techniques. For, as we continue with our discussion of work sampling, it should be kept in mind that we are primarily referring to it as a tool useful at a certain stage of a cost improvement program and not as a universally applicable technique to solve every problem. No technique can be classified in this way.

HOW TO TAKE A WORK SAMPLING STUDY

Part One dealt with cost improvement programs; a brief discussion of work sampling was given, with emphasis on the fact that work sampling has proved to be a very useful technique for analysis where either personnel costs or equipment utilization were crucial to a work situation. In other words, when the initial examination showed a large proportion of the budget in either labor or equipment costs, the next step suggested was that a work sampling study be conducted to analyze these expenses. Also, from time to time throughout Part One, references were made to various activities, such as the selection of measures of output, that are useful not only in the first phases of a cost improvement program but are also vital to the proper conduct of a work sampling study. These comments may be referred to in our discussion of work sampling, but will not be repeated. As a matter of fact, one reason for the approach given for the operation of a cost improvement program is that such a program provides a very solid foundation for the following steps. However, although our discussion of work sampling will not be entirely self-sufficient, its scope will extend well beyond the use of the work sampling as part of a formal cost improvement program.

The objective of the discussion that follows is to describe the procedure for taking a work sampling study. Generally speaking, the steps described will be necessary for all studies. This part of the book should serve as a "how-to" guide to work sampling. (A previous book, *Work Sampling*, co-authored with Robert E. Heiland, is a more comprehensive treatment of work sampling.) Our purpose here is to reflect changes that have occurred over the past years, and to give the reader sufficient instruction so that he can carry out work sampling studies, for whatever purpose. We shall therefore start by discussing each step in some detail, and conclude with an example, which should give a coherent picture.

Setting the Objective of the Study

Work sampling is a versatile technique and can be used to accomplish many objectives. But before we start a study, it is important to define specifically the end uses to which we intend to put the study. So many different variations are possible in our selection of work units, observers, and patterns of observation that it becomes necessary to specify each of these in the light of what we expect to get from the study. To help the reader decide upon his objective, a sensible approach might be to give some possible uses of work sampling so that he can see where his own needs fit in. These uses will be given in the order of the greater number of applications, as the author has experienced them. They can generally be classified as follows:

Use of work sampling as a primary analysis tool for an indirect and service activity which is largely unmeasured. By far the largest number of applicatons of work sampling have occurred in areas such as maintenance, clerical, and service activities. It is characteristic in most cases that it is the first attempt at analysis or measurement. It is also characteristic that such studies are part of a management effort to extend work measurement to these areas, which, again, are largely unmeasured. Management's desire is to take the necessary first step in establishing what the work pattern is now in these areas. A few broad categories are used, and large numbers of observations are not necessary. Basically, we want to find out what our people are doing with their time, so that we can take appropriate action toward improvement or simply the installation of standard labor control procedures. The example already given, of the maintenance department, is typical. Cost improvement in this instance is more likely to come from better planning and scheduling than from an intensive effort in methods work.

This type of study, basically a fact-finding preliminary to management action, is by far the most common. In addition, this study is used in most cases as a bench mark against which to measure change. The advantage of such a study is that it measures directly the effect of our management policies and controls. As an example, we may have a large backlog of work in some area and still find an inordinate percentage of waiting, delay, and personal time. Finally, it should be emphasized that this type of fact-finding study is most appropriate for initial attempts to bring under management control every type of indirect work which previously have been regarded as too nonrepetitive to be measured.

Use of work sampling as a technique in supervisory development. A basic problem facing all management is the development of first-line supervision into effective managers. There is a constant turnover of first-line supervisors, and management has long recognized the need to provide a systematic program to develop supervisory skills at the first-line level. The active participation by the supervisor in a work sampling study of his own area is a most effective way in which to further his development. First, his participation will be real, and will be seen by the supervisor as more than just a training exercise. Second, participation in the program as a sampler will in fact give him a much better insight into just what his people or equipment are doing. Third, the combination of the work sampling for input and work units of output will make clear to him the basic objective of his department. Fourth, there is the advantage that he is learning a new skill and learning about methods and standards in a situation which he feels is only partly personal development and mainly the accomplishment of a genuine management task. Finally, the supervisor as observer is much more likely to understand the results and to cooperate in working toward the objective of the study.

For work sampling studies made of indirect or service activity, the supervisor is indeed the most logical person to take the observations. He knows the work and he knows the people. It is much easier to train a supervisor to take a work sampling than it is to assign an industrial engineer to the area where he must learn about the personnel and work practices in that area. Whatever time is saved because the industrial engineer is skilled in work sampling may be lost in the time necessary for him to become familiar enough with the work situation to be an effective sampler. The supervisor may protest that "he does not have enough time" to do the sampling. There are two answers to this: (1) he should realize that part of his job is to be out on the floor where the work is being done, so he will be there part-time anyway, and (2) he will probably prefer to do the observations himself rather than have a stranger introduced into his work area.

An objection sometimes raised is that the supervisor will not be unbiased in his observations; he will slant his observations so that his operation will "look good." This does not seem to happen very often. In practice, the first work sampling study usually serves as a bench mark for his department, and he will probably be perceptive enough to realize that the pattern of his work sampling should be reflected in the work units of output. In other words, he will be keeping records of the output of his department, and will match these with the input of work as shown by the work sampling. If his people are invariably productive, the input-output relationship will be quite straightforward. If, on the other hand, delays and lost motion occur but are not reflected in the work sample, irregularities in the input-output relationship will be readily apparent. But there is an even more rational answer to what is perceived by some as a problem. Our first-line supervisors are usually effective employees. If the study is explained properly and we do in fact genuinely make them a part of our effort, the results will not be disappointing. Every example that will be given, with one exception, was taken with supervisors acting as observers.

Studies made to establish time-study allowances. When standards are set on repetitive work either with a stopwatch or a human motion-time system, they cover only the repetitive part of the job. The entire job will also include various miscellaneous delays, delays for tool care, delays for material handling, and other normal delays of the work environment. It was stated previously that these delays are covered by allowances, or flat percentage adjustments, which add the necessary time to the standards covering repetitive work. It must be understood that such allowances are shopwide because the conditions usually exist throughout the shop. It must also be understood that if there is in fact a legitimate cause for delay, the people who are doing the work must be given time to compensate for this, and the time must be in their standards. In all too many cases, these allowances are negotiated, estimated, or made the subject of recurring traditional practices. Since such allowances might amount to 10 or 15 percent, it is quite apparent that they really should be determined in a fairly systematic manner. For this reason, work sampling has come to be used quite frequently for allowances for repetitive work.

The obvious first step in making such a study is to be sure that the categories of activity be defined quite carefully to include specific causes for delay, and to exclude work normally covered with the stopwatch or predetermined human motion-time standards. Also, the industrial engineer may take these studies, because they are usually considered part of the work measurement process. It is by no means unusual to have the supervisor take a few samples in order to satisfy himself about the reliability of the percentages arrived at. In such a study, a fairly large number of observations is usually sought, because some precision is required. On the other hand, since allowances usually are shop-wide, the population sampled is usually large, and the obtaining of large numbers of observations poses no real problem.

Identification of cycles within a work month. Work sampling is sometimes used to determine the presence or absence of cycles in the pattern of work. Usually, work sampling studies are taken within a one-month time period. Because of sample-size limitations, daily figures probably will not prove to be sufficiently accurate. A common practice is to express sampling percentages in terms of weekly figures. This seems to be a reasonable compromise between the variability of daily figures and the lack of sensitivity of monthly figures. The results of the work sampling should, of course, be matched with the records of output, and the comparisons made on the basis of unit times. Even here, there probably will be no clear-cut resolution. But it is useful to make comparisons of the activity pattern from week to week. One can always add the weeks into a month. A word is also in order here about "cumulative" results. The practice of adding each day successively to the total until "stability" has been achieved is, in the author's opinion, a form of self-deception. As the sample size increases in total, the effect of any short period is submerged within this total. Thus, far from "settling down," the situation may be just as variable as ever, but this variability is not reflected in a cumulative figure.

Special work sampling studies. Sometimes work sampling studies are conducted in order to analyze a particular pattern of employee activity. For example, a study of nurses included a special category for "formal training" so that hospital management could see whether a proper amount of time was being devoted during an eight-hour shift on a hospital floor to the training requirements for probationary nurses. As another example, a special category was put into a study of a group of welders to see whether or not welding machines were being used where possible. This study was made by the supervisor, and an element of judgment entered into the observations. But however implemented, work sampling has proved to be a very useful tool in answering specific management questions about actual operations. These studies are not limited to wage-roll employees. In one particular case to be studied in detail later, foremen were sampled to determine what percentage of their time was spent out with their employees, at the scene where work was being done. No attempt was made to appraise the level of effectiveness of the foreman as he did his job; the only question asked was whether or not he was at the work station.

In summary, there are many uses of work sampling studies, only a few of which have been enumerated here. But the uses given are the most important, and the reader should be able to extend these to include any particular objective. The important point is that before a study is started a full discussion should be had on the eventual uses to which the study will be put. These should be written to serve as guidance in the technical work of planning and carrying out the study. They also will be invaluable in explaining to the employees, management, and union the exact nature of the study.

Defining the Population to Be Studied

Whenever we do any kind of sampling, we select certain items in a population, draw the sample in terms of the activities of these items, and extend the items of the sample to include the entire population, which may include personnel or equipment or both. In our situation, we are going to sample the activities of people or machines during part of their work cycle, and draw inferences about the total work cycle and the total number of people or machines from our sample. It is therefore important that we define exactly which people we are going to sample, or which machines. If we are going to sample during the first shift only, we should go over the organizational structure, decide which people are to be included, and then go over the payroll records to be sure that the population we are sampling can be isolated in terms of total hours. We must also be sure that we have included in our sample personnel or machines that match the organizational structure of accountability and for whom we are collecting work units of output. This should present no problems, particularly if the sampling is being done in connection with an organized cost improvement program. In such programs, the first step is to establish budget accountability, which is usually done along straight organizational lines. We also establish our work units of output, and this too is generally done following the existing organizational pattern.

In specifying the exact makeup of the sample, a very common condition occurs when we have a large group of people or machines who perform roughly the same function, and we thus do not feel it practical to sample the entire group. We then must select subgroups, probably of the size supervised by one first-line supervisor. It is important that these subgroups be representative of the entire activity. For example, in one work sampling done at a large construction site there were over 200 men in the electrical craft. These were organized so that a dozen or so worked under the direction of each foreman. Because the foremen were members of the bargaining unit and could not act as samplers, salaried employees were used.

Because the number of samplers was thus limited, it was decided to sample four different groups of electricians, each group doing typical electrician's work. To have complete coverage of all the different kinds of work that electricians do, one group was chosen that was doing conduit work, another group doing instrument work, a third group doing wiring work in cable support trays, and a fourth group doing panel wiring. It was felt that taken together these four groups presented a cross section of the total effort. If for some reason it is necessary to select subgroups such as these from a large overall population, the commonsense notion that the group should be representative in proportion to the total work distribution should apply.

Sometimes it is desired to sample a group of people or machines that is quite widely dispersed in the geographical sense. In this case, areas sometimes are sampled as geographical zones, and activity in each zone is assumed to be normal. This may be a dangerous assumption, particularly when one is sampling people, because what does not show up is personal time and time off the site. Also, it is quite difficult to keep in mind a large number of employees, and we cannot honestly define the population from which we are sampling. The recommendation in this case is that the group be defined, and each member of the group be included in the sample. If a member cannot be found after a reasonable effort, he is simply noted under a "no contact" category. This practice of attempting to get a sampling by individuals will result in a smaller number of observations. But the observations we do get will be much more meaningful.

In summary, we usually sample groups that are clearly identifiable on the organization chart, that are supervised by one person, who is accountable for their activity, and that have work units of output which are identifiable from normal accounting records. In the opinion of the author, it is better to accept a smaller number of observations for a well-defined group than a larger number of observations taken in a group whose composition is not clearly defined.

Selecting Measures of Output to Reflect the Activity of the People or Machines

This topic has been covered in great detail in our previous discussion of cost improvement programs. But it is of such basic importance that it should be mentioned here again. Work sampling is a measure of the distribution of time or of the input to a work situation. The conditions under which the work sampling study is taken are noted, but not specified in detail. If we are anxious to extend any inferences drawn from the study, we must at least know what output was produced during the period of time over which we used work sampling to analyze the input. This is the basic reason for the author's recommendation that the work sampling be done over a period of one calendar month or one accounting month. Since we usually keep records of output on at least a monthly basis, we will be able to draw inferences concerning the level of activity in the areas sampled by comparing the output for the period sampled with the historical pattern of output.

A previous example gave historical data in terms of the monthly figures for maintenance work orders closed out, which is a common measure in maintenance. If we do not decide upon and keep track of our records of output, we will not be able to extend the results of our study very far. A very valid criticism might be that in the sampled month business was "unusually slow." Or "we had an unusually large number of small orders." Or "we were doing a lot of special work that month." It may be true that the month sampled was unusual, but we should at least have records to support the exact conditions.

Generally speaking, no new records will be required to keep track of the output for the period for which the work sampling is done. After all, we do keep records now of the work that each function performs. We do this because the work units are either sold to customers, transferred to another department, or represent a service that usually is made the subject of some accountability. Sometimes we will ask a supervisor to pay particular attention to a classification of work unit, but the usual case is that the records we want are already available. It is simply a matter of deciding which measures most fairly represent the output of the group sampled.

It is also worthwhile to note that the recording of work units of output and the work time of input are necessary for the development of some of the statistical measurements to which we have referred. If we have decided that we will keep these input-output data, we should make as extensive use of them as possible. So a month is a fairly reasonable time period for the collection of data; we might be able to combine the work sampling, the work units, and a statistical or mathematical analysis so that we have a real bench mark, which can be used with a minimum of uncertainty.

Finally, in the case of indirect work, it has been the author's observation that too often control is exercised simply by looking at input, or labor cost. The real basis for improvement usually comes when we look critically at output, particularly as we ask the question, "Is this work unit really worthwhile?" By systematically examining the work units of output we can bring to management's attention the specific return for each function. Whether this meets the current objectives of management is a basic question of managerial judgment. But, at the very least, the question will be asked and the judgment exercised.

Defining the Period of Time over Which the Sampling Will be Done

Whenever we sample, we must define exactly the population from which the sample is drawn. This definition should specify both short- and long-term periods. We have already discussed the need for exact specification of which people or groups should be sampled. We must also specify exactly which time periods constitute the population from which we sample. We must do this first in the long term, that is, which shift and which calendar day will constitute the population. But for two basic reasons it is equally important to specify exactly the minutes of the day during which the sample will be drawn: (1) It is necessary to know the specific daily pattern in order that we may set up random numbers which represent sampling times, and also to be able to convert the percentages of observations into hours of time and thus into cost. In other words, we need to specify exactly the daily schedule, so that the mechanics of taking the sample and converting the results to time and costs can be accomplished. (2) We should specify exactly the population we are going to sample during the day, because there will be periods of time during which we probably will not sample; these should be identified and excluded from our population. For example, we do not normally sample during regularly scheduled rest periods or personal time breaks. This seems only sensible, because sampling during these periods would probably be regarded as harassment by the employees, and in fact would not give us any useful information that we did not already know. On the other hand, if employees are allowed a certain amount of personal time, to be taken at the employees' convenience, we would probably include the entire day in our sample and then project the result to reflect this allowed time.

In addition to personal time, there are other classes of activity by the employees that we may want to exclude from the sample. For example, the first five minutes of work in some maintenance operations is assumed to be taken up with either traveling to the job or preparing the site for work. We sometimes do not include this time in the sample, but rather arbitrarily define these parts of the working day as being devoted to preparation or travel. In such situations, we end up sampling only time during which the employees are supposed to be productively employed. Figure 14-14 is an example of such an arrangement, and includes the results that came from the sampling.

For the specification of the short-term sampling period, our best guide is the stated company policy or the union-management agreement. Many instances exist in which, over the years, special allowances given for temporary situations become permanent. It is, of course, very difficult to "turn back the clock" if these circumstances now seem to favor the employee. But at the very least we should govern our sampling population by either company policy or the labor contract or both. Table 14-2 is an example of a statement of both company policy and the conditions set forth in the contract. This was used to establish the short-term time periods over which the sampling was done. Again, the purpose of this specific detailing of the times that form our population for sampling is essential not only to extend the usefulness of the study and draw valid inferences, but also as a means of reviewing actual practice and its relationship to stated policy.

One final example may be illuminating. In the maintenance operation of a large oil refinery, which the author considered to be typically well managed, this particular step of defining the short-term population produced an interesting situation. When the maintenance supervisors for the various crafts got together to perform this step, the fact emerged that what seemed to be a clear-cut statement of the work day was in fact not quite so clear-cut. The contract and company rules stated that "The work day will begin at 7:30." However, one supervisor said that in his craft this meant that the men were at the tool box (near the work site) with their protective equipment and hand tools, ready to go to work. Another supervisor said that in his craft this rule meant 7:30 in the

Rather than face the argument involved in checking a large group traveling or at rest, during break we stopped the trip five minutes before a scheduled break until five minutes after. The same goes for lunch and quitting time. This time was deducted resulting in samples of a 6.5 hour day rather than 8.0. To return the results to an 8.0 hour day the following adjustments were made:

8:00 to 8:05	5 Min. added to work preparation
9:25 to 9:45	10 Min. added to travel
	10 Min. added to personal (break)
11:55 to 1:05	10 Min. added to travel
	15 Min. added to personal (wash up time)
2:55 to 3:15	10 Min. added to travel
	10 Min. added to personal (break)
4:25 to 4:45	5 Min. added to travel
	15 Min. added to personal (wash up time)

This tallies out to a total adjustment of:

Work Preparation	0.08 hrs.
Traveling	0.58 hrs.
Personal	0.84 hrs.

The adjusted percentages were determined as follows:

	%	*Hrs.*	*Adjust*	*Total*	*Adjusted*
Direct Work	47.9	3.12		3.12	39.0 ⎫
Work Preparation	22.8	1.48	+ .08	1.56	19.5 ⎬ 58.5%
Travel	15.6	1.02	+ .58	1.60	20.0 ⎭
Unavoidable Delay	2.2	.14		.14	1.8
Personal	2.8	.18	+ .84	1.02	12.7
No Contact	8.7	.56		.56	7.0
	100.0	6.5	1.5	8.0	100.0

Figure 14-14.

change house, with the craftsman getting dressed to go out to the work site. Finally, one supervisor said that in his craft it meant 7:30 inside the main gate, with the employee's time card punched in. The problem presented by these three interpretations of what seems like a straightforward work rule is obvious. But no matter what the resolution of this problem, the sampling day has to be defined in quite specific terms.

Table 14-2.   WORK SCHEDULE
MAINTENANCE AND SALVAGE

Item	Contractual	Our Operation
Starting Time	Between 6:00 and 8:30 a.m. or 2:30 and 4:00 p.m.	8:00 a.m. for the day shift and 3:30 p.m. for evening shift, unless otherwise scheduled.
Morning Rest Period	10 minutes during first four hours of any shift.	A maximum of 15 minutes is allowed when walking from and to distant jobs. Times as posted for various shops.
Lunch Period	30 minutes plus five minutes wash-up.	35 minutes off the job—11:40 a.m. to 12:15 p.m. for the day shift. 7:55 p.m. to 8:30 p.m. for the evening shift.
Afternoon	None	Same as morning rest period.
Quitting Time	Eight hours from starting time, exclusive of lunch periods, less 15 minutes shower time.	Additional time is required for putting away tools and preparing time cards. For the day shift, a man should not enter the Shop area from a job that is not complete before 4:05. A man cannot leave the shop area for the showers before the bell rings or 4:13 for the Salvage Department. For the evening shift, a man cannot leave the Shop area before 11:43.

Selection of the People Who Will Do the Sampling

This aspect of the study has been discussed previously, since work sampling, like most work measurement techniques, can accomplish several objectives at the same time. It is particularly appropriate that the first-line supervisor be used to take the samples in indirect work situations. It should be pointed out here that the distinction between "direct" and "indirect" work is really a measurement or accounting distinction. Where the work is sufficiently repetitive in nature, we establish the input-output relationship in terms of a standard time per work unit. We then charge that standard time against each work unit and apply the cost directly to the work unit produced. In most cases where the work is nonrepetitive, or where it would be too difficult or expensive to measure, we simply throw the cost of such work units into a pool called overhead or factory expense. Such functions as maintenance, supervision, materials handling, and shop clerical work are commonly charged this way. The cost of such work is then redistributed over the product or service produced on the basis of man-hours, labor dollars, or some other scheme. In management's drive to do a more precise job of distributing this cost, work sampling is frequently the first technique to be applied. What makes it particularly appropriate to use the supervisor as the observer in such cases is that he knows the work and he knows the people. He can easily be taught to take the observations.

In almost every instance, but particularly in the case of an organized cost improvement program, the cooperation and active support of the first-line supervisor in analyzing the work and introducing change are essential. Also, the first work sampling study in a work situation where very little standardization, planning, scheduling, or methods work has been done will almost always show a relatively modest amount of productive work. This will please neither the supervisor nor his manager. The first impulse would be to deny the reliability of any study which showed, for example, that maintenance people are engaged in productive activity from only 35 to 40 percent of the time. (This is not an "ironclad" figure, but is quite common from studies taken as a first attempt to exercise maintenance control.) If the supervisor has participated in taking the observations, and if he is guaranteed that his "throat will not be cut" as a result of this first study, it has been the author's experience that the supervisor makes a very satisfactory observer.

One aspect of having the supervisor take the observations is that his very presence moving around and observing his people may alter somewhat the work situation as it exists. This is no doubt true, but it is not necessarily a bad thing. After all, we do want the supervisor to be shown at his best. And it is a common supervisory complaint that the demands of "paperwork" and other management programs too often occupy an excessive amount of his attention. In one plant, for example, eleven special management programs were to be "given priority" by the supervisor. If management is serious about cost improvement, the supervisor's priorities will include the time to take the work sampling studies, and it is hard to see what is more useful than this requirement, which really makes it possible for the supervisor to get a systematic analysis of his people and equipment. Most other programs start with this information.

A final word is in order concerning the use of the supervisor as an observer in the case where an organized cost improvement program is being pursued. The work sampling study will demonstrate that the supervisor does in fact have some time which can be made available for analysis and improvement of his operation. The work sampling is the analysis phase, and will undoubtedly reveal specific areas requiring further attention by the supervisor. When the work sampling has been finished, we simply make the point that since the supervisor has had time to do the sampling, he also can make time to work on a specific improvement.

The supervisor seems to be the logical person to do the sampling in indirect areas, or where a regularly assigned industrial engineer is not available. In the production area, however, industrial engineers usually are available. Here the industrial engineer may be the most appropriate observer, particularly when an objective of the study is to set time-study allowances. The supervisor can be invited to take some observations to satisfy himself of the reliability of the study, but in most operations the supervisor does not regard it as his responsibility to become involved in any form of work measurement. If an experienced industrial engineer is available, who has been working in the particular area, he is a very logical choice to serve as observer. He probably knows the work and the people; in addition, he should possess the appropriate skills to organize and conduct the study.

There are a few special cases in which observers other than the supervisors may be used. For example, when the supervisor is part of the bargaining unit, as is quite common in the construction crafts, it is almost mandatory that someone not in the craft be used as an observer. A very common solution here is to train an engineer who is familiar with the work to act as observer. In several cases, mechanical, electrical, or civil engineers already working on a construction site have been trained and used as observers. Training such people is quite simple and can be done in a few days. It then usually requires a week or two of exposure to the particular area and group so that the observers will learn to recognize by sight the men to be sampled, and also to learn the details of the job and the scheduling system in effect. Some of this they may already know. But it may not be fair to ask an individual who is serving as a craft foreman on one job, but who on the next job will be a journeyman, working for some other member of the bargaining group, to act as observer. Most trade unions reserve the right to challenge work measurement procedures, but do not want to have their members actively involved in such procedures. This is a viewpoint with which the author has a great deal of sympathy.

When observers have been selected, they should be trained in the technique of work sampling. Usually, this training consists of three or four half-day sessions, followed by on-site training by the person responsible for the overall program. The formal training varies with the experience of the observer. However, the first session usually covers the basic technique and the objectives of the study. Use of film loops and a minimum exposure to the statistics involved usually are features of this session. The observers are then asked to define the group sampled, the work day, and the measures of output. They are allowed about a week to do this.

In the second session, a set of categories of activities and random times will be developed. Then the people who are to be studied should be informed, and specific tests made of the validity of the population to be sampled and the work units to be used. This again takes about a week.

The third session is a preliminary to a two- or three-day test sampling. All questions relating to the sampling can be covered here. Then sometime during the following week, usually, a three-day trial sample is taken.

The fourth training session involves discussion of the trial sample and the making of decisions to resolve whatever problems have arisen. At this point we are ready to continue with the regular work sampling. It is important to note that the training requires only a minimum of the observer's time, but is conducted over a calendar period of three or four weeks. It is also important to note that the entire organization and all supervisors are not usually made the subject of the first work sampling. It is much more common to start actively sampling with three or four supervisors, even though others may be trained at the same time. This has been discussed previously.

In summary, then, a choice usually exists as to who should act as observer. In indirect areas, there may be compelling reasons to use a supervisor. But this is very much a management decision. The quality of the observations is, after all, the foundation of any study. The mechanics of taking observations are not very complicated, and the mental attitude and experience of the observer are at least as important as the technical skill necessary to do the job.

Formulating Activity Categories for Sampling

In work sampling, the population from which we sample is the total activity of a precisely specified group of people or machines over a precisely specified period of time. The "total activity" to which we refer consists of everything the people or machines do during the time period. If we were making a narrative account of this activity, we would start perhaps with an operator's setting up a job, getting material, walking to the tool crib, walking back, and so forth, throughout the entire day. What we would have for our expenditure of a man-day's observer time would be a good analysis of one person's activity. Instead of this, we divide the total activity of those people or machines observed into classifications of activity, which we call categories. These categories group together similar activities and should reflect the objectives of the study. They should be designed for ease of observation and consistency with the objectives of the study.

For example, if we were sampling the activity of nurses on a hospital floor, we might take all instances in which a nurse is attending directly to the needs of the patient and define these as forming one category of "patient care." Suppose that these instances included feeding the patient, bathing the patient, taking the patient's temperature, or simply talking with the patient. We would lose our ability to determine exactly how much of the nurse's effort is spent in feeding the patient as opposed to the other activities, but we would have a broad, easily classified series of activities, which, if our objective were to establish the extent of the nurse's involvement with the patient, would be suitable. On the other hand, if one objective in our study was to find out how much time nurses spent in the process of delivering food to the patients, we would need a separate category for this.

In a similar fashion, the activities of equipment can be reduced to categories. An example is shown in Figure 14-15 where a large number of punch presses were sampled. The objective was to find out exactly why the presses were not producing at rated capacity. The categories reflect the various activities, or lack of activity, that are normally found in a press room. In this particular case, it is obvious that further questions had to be asked after some of the observations in order to establish the reason for a particular activity.

As a general rule, for the most common type of study, taken as the first detailed analysis of a work situation, it is good practice to have a relatively small number of categories. Usually six or seven are sufficient, and if the number is held below ten, a single digit will suffice for recording the observations. Also, the temptation to solve all the problems in one study should be resisted. Pareto's law works here, as elsewhere. In a typical study with ten categories, the first three or four usually account for almost all the pertinent activity. The example given at the start of the discussion of work sampling is quite representative, even to the distribution of time and proportions of observations. Specifically, categories 1, "travel," 4, "work," and 6, "personal," amount to about 70 percent of the observations, even though they are only three of the eight categories. There may be special cases in which some particular activity is important, but regardless of its magnitude, the general rule in developing categories is to keep them simple and not to have too many.

The next point, that the categories be written, also is illustrated by the same example given earlier. The definitions of the categories should be agreed upon, not only for the obvious reason that this is necessary for the training of observers, but also in order that they be capable of being checked against objectives. When future studies are desired to check progress in the same situations, the same categories may be used.

A very useful category of activity may be that of "no contact." When the people to be observed may be found in different locations, such as is the case in maintenance, we must be ready to accept the fact that we may not observe a particular person on a particular round of observations. The person who is to be observed may be walking down one side of a building while the observer walks up the other, he may be in a restroom, or he may have gone on an errand. We do not want our work sampling study to become a gestapo-like operation, so we usually do not sample the restrooms; if we miss a person, we make the only judgment we can, that we did not observe him, or "no contact."

One advantage of having the supervisor doing the sampling is that he will have a much better understanding of where his people are or might be. But the

AMP 1838 Summary of Work Sampling	Observer *IVAN KERNS*	Dept. #	Period Covered	No. of Employees
Cat. No.	Category		Number of observations	% of Time
	PRESSES — PIKE PLANT			
	PRESS ROOM (44 PRESSES)			
1	*RUNNING*		*807*	*36.7*
2	*NOT RUNNING - FULL REEL*		*151*	*6.9*
3	*" " ORDER COMPLETE - OR NO DIE IN PRESS*		*140*	*6.4*
4	*" " OUT OF SPEC.*		*27*	*1.2*
5	*" " DIE REPAIR*		*621*	*28.2*
6	*" " PRESS MAINT.*		*22*	*1.0*
7	*" " AUTOMATIC*		*51*	*2.3*
8	*" " JAMMING*		*26*	*1.2*
9	*" " OUT OF STOCK*		*36*	*1.3*
10	*" " SET - UP*		*72*	*3.3*
11	*" " OPERATOR WORKING AT PRESS*		*59*	*2.7*
12	*" " OTHER*		*188*	*8.5*
		Total	*2200*	*100*

Figure 14-15.

inclusion of a "no contact" category not only reduces the times or tours of observations, but also makes them seem more reasonable for the supervisor to take and to the employee who is being observed. Some of the "no contact" will be legitimate personal time, no doubt. But the supervisor and the director of the study will be able together to make an assessment of the situation. In any event, the inferences we draw from the category will be governed by the proportions of observations that fall into it.

One pitfall to be avoided is the definition of categories to include one category of "work" or "productive activity" and to have all the rest obviously of a nonproductive nature. We must remember that in most jobs there is auxiliary activity necessary to the work. Also, we are using the categorizing as a fact-finding device, and we should take advantage of its positive aspects. This is not as important in studies of machine utilization as it is in studies of people.

In cases where we are sampling jobs with very well defined physical activity—such as maintenance, nursing, materials handling, and inspection—the task of classifying observed activities into categories is fairly straightforward. By that we mean that the observer can make the judgment accurately without interrupting the activity of the person observed. There are other situations in which observations must be made in two parts. We want to know not only what the person is doing but why.

For example, a group of engineers in a chemical product development group had a work pattern that included talking, calculating, and telephoning, not only for the product on which they currently were working, but also including inquiries concerning products that had passed from the development stage into the production stage. The observer was able to tell what the engineer was doing, but it was desired to associate that activity with the particular product. This was necessary to establish what development effort went into products that had already gone through the formal development process. In this case, a small card was laid out with a double-spaced list of six products in which there was an interest. An additional item labeled "other activity" completed the list. When the observer walked by he first made his observation of what the engineer was doing, then nodded to the engineer, who would put his finger either on one of the products or on the notation "other activity." Needless to say, this system of taking the second part of the observation was evolved with the full cooperation of the engineers. All sorts of variations such as this are possible.

Another unusual situation, which has been discussed already, was that of sampling foremen in a construction activity. These foremen, members of the union, were working as part of a contractual agreement. They were sampled for two categories only: "at the job site" and "not at the job site." The company took the position that since all planning and scheduling work was supposed to be done by other employees, the foremen should be at the job site. This may seem to be a minimal objective for a study, but it did serve a special objective.

One other special type of category might be mentioned. We sometimes wish to associate activity with a particular kind of work unit. That is, we want to find out the position of productive activity among several kinds of work units. To do this we use two sets of categories. The first is a normal set of categories of activities and the second a set of categories of work units. This usually occurs in a second study, for we try to keep our first study as simple as possible.

In summary, we divide the total activity that is our population into sampling classifications, which we call categories. Each category is assigned a number, and the observations made and recorded as the number of the appropriate category. No one set of categories will cover every situation. As a matter of fact, categories should be devised to meet the particular object of a study, should be written, and where possible should be capable of association with the work units of output. In this section and throughout the discussion of work sampling, several examples are given of sets of categories. But these should really be developed locally through discussion with the observers.

Deciding upon the Number of Observations to Be Sought

Practical aspects of the sampling situation should be the controlling factor in determining the number of observations, and the imposing of artificial requirements of reliability should be avoided. Work sampling, of course, is like any other sampling procedure in that we draw inferences concerning the entire population by considering only a part of that population. As is the case with all sampling, there is some uncertainty in our statement of the inferences that we draw. It is only normal and natural, and since sampling procedures play so large a part in measurement and indeed in management, the concept that there is uncertainty is not a strange one. Moreover, there is a body of knowledge about sampling that will allow us to make a quantitative statement about the uncertainty of our results. Uncertainty can be caused in two ways. First is uncertainty because of the size of the sample or the number of items that we draw. It is obvious that the larger the sample, the less the uncertainty. We shall pursue this concept in our discussion of the reliability of the sample. Second is uncertainty due to error in the actual classification of the items sampled, or due to the fact that the items sampled may not thoroughly represent the population. In general terms, we refer to this as the validity of the study.

Let us first discuss the matter of sample size. Before embarking upon the mathematical rules that govern sampling, let us first put in perspective the practical constraints that govern the work sampling study. We must remember that, regardless of the sample size we would like to have, the sample which we are able to get depends upon the product of the following four constraints: (1) the size of the group sampled, (2) the number of observers, (3) the number of rounds per day that the observer is willing and able to take (an observation is one judgment made of the person or machine; a round of observations is the process of making this judgment for each person or machine in the group), and (4) the number of days over which the study is taken.

Let us take these in order. It seems a very obvious statement that we are constrained by the number of people in a group. Yet we have already made the point that this group must be defined quite precisely; thus, if we have five mechanics in a group, there is a constraint. We are quite unlikely to add another five for no other purpose than to increase the sample size. The number of observations is therefore limited to those we can take on the five mechanics.

The next constraint is the number of observers. In most cases a single observer is used. Sometimes, when a foreman is faced with obligations that prevent him from making his round, an engineer may fill in as observer. Sometimes, in large groups, we use two observers. But an overwhelming majority of work sampling studies are taken with one observer. This is particularly true of studies taken as part of a cost improvement program.

The number of rounds per day that the observer is willing and able to take is more of a variable. In some types of studies, owing to the geography of the sampling site, the number may be quite limited. In maintenance work, for example, as few as three or four rounds per day may be a reasonable number. And when the first-line supervisor is taking observations, we must remember that we have assigned him this as a collateral duty; he still has other work to do. In studies taken in a limited geographical area, it would be possible to make perhaps eight or twelve rounds a day. In any event, the number of rounds per day is usually arrived at by a form of bargaining with the supervisor who is taking observations. Above all, we are anxious to avoid a careless approach to the taking of observations. Four rounds a day from an observer who conscientiously does a good job is preferable to eight rounds a day by an observer who accepts his task as "extra duty" and is not conscientious about the procedures he is supposed to follow as an observer.

Finally, the last constraint is the number of days over which the study is to be taken. If we follow the practice of taking a study over a calendar month, the constraint will be the working days in that month. (There are good reasons for sampling over a month. The work units of output are usually kept on a monthly basis, within a one-month period a cycle can be detected, and a monthly sample usually gives us a reasonable number of observations.) So this final constraint is a very practical one.

Let us take as an example a group of ten maintenance craftsman, all pipe fitters. Let us assume that their work will take them into fairly widely dispersed locations and that the supervisor has agreed to take the observations, making four rounds a day. Finally, let us say that there are twenty-two working days in the month to be studied. The arithmetic of this situation is that the total number of observations equals 10 men sampled $\times$ 1 observer $\times$ 4 rounds per day $\times$ 22 days, or 880 observations. This is the practical limit of sample size. We can alter the number of observers, the rounds per day, or the days in the study, but particularly for our first work sampling study of a fact-finding nature this may lead to problems. Specifically, the supervisor may feel that even four rounds a day are too much of a drain on his time, but he has agreed to do a good job with those four rounds. It is possible to state the sampling error in quantitative terms. Then the decision must be made as to whether or not this error is acceptable.

The last point is important. With the exception of work sampling studies taken on a shopwide basis for the setting of time-study allowances, the author has little sympathy with blanket requirements for reliability based upon work sampling studies. This does not mean that reliability is not an important factor in a study. But the number of observations that can be made depends upon the given constraints. A corollary of this is that blanket rules for work sampling studies do not take into account the constraints previously mentioned. A blanket rule ignores the size of the group, the number of rounds per day, the number of observers, and the number of days in the study. It simply does not make sense to try to apply the same standards of reliability to a two-man group as to a twenty-man group. It is also, practically speaking, not very sensible to insist on the same number of rounds per day from a supervisor who is in a centralized clerical function as we do from a supervisor whose people are dispersed geographically. Finally, we should look to the objectives of the study and the uses to which the results will be put. Again, it does not make much sense to apply the same rules to a shopwide study for the purposes of determining allowances as it does for an exploratory study of a small group for the initial analysis and determination of the direction of improvement.

The preceding does not mean that we should disregard the question of reliability. But the reader should be reassured that the problem sometimes tends to solve itself. For example, a shopwide study, made by an industrial engineer for the purpose of determining allowances in time standards, is apt to cover a large group of people or machines. Often the industrial engineer is a skilled observer. Because of the large number of people involved the industrial engineer can probably make a substantial number of rounds of observations. The total number of observations is also apt to be large enough that we can apply certain requirements of reliability. To continue this thought, when we are dealing with larger groups of people, where the personnel costs are quite significant, we automatically get a larger number of observations. Thus we will be more certain of the result and will take any action that seems appropriate with a great deal more confidence.

In summary, for this part of the discussion, the point has been made that there are practical limitations on the number of observations which can in fact be obtained in any given situation. It is often true that we would like to have many more observations, just as we would like to be more certain of many other management figures that we use. However, the limitations are there, and we should recognize them as a fact of life. This seems to be a more intelligent approach than to "solve" an algorithm to tell us the number of observations we "should" have.

It now seems appropriate to have a brief discussion of the theory of work sampling. First, we should list the assumptions that we make. We assume that although the sampling situation is really a multinomial, that is, a sampling of several categories, we can safely treat the sampling situation as if it were a binomial. By this we mean that we have just two categories: the one in which we have an interest, and one composed of all the others lumped together. The theory of the binomial distribution holds that the proportion, p, of a particular category is held to remain constant within the population; that is, taking the sample does not alter the basic proportion because the population is held to be infinite. This proportion, p, is of course our chief interest. We are concerned with stating the reliability of the value p in any sample that we take. Common sense tells us that the larger the sample, the less likelihood there is of sampling error. Therefore, when we calculate an expression for the sampling error (σ) of p, we would expect to find the sample size, N, in the denominator. In fact, the mathematical expression for σ_p is as follows:

$$\sigma_p = \sqrt{\frac{p(1-p)}{N}}$$

Disregarding the numerator for a moment, we see that this error will decrease as N increases. However, the magnitude of this error does not vary inversely with N, but rather inversely with the square root of N. Thus we will see improvement in the form of a reduction in σ_p as the size increases; but since we are considering the square root of the number we soon reach a point of diminishing returns. For example, the difference between the square root of 100 and 1000 is substantial; however, the difference between the square root of 1000 and of 2000 is not so substantial. And σ_p, it must be remembered, is a measure of dispersion or variability. It gives us some idea, in quantitative terms, of the magnitude of a standard error of measurement.

Any statement of uncertainty should really be made in two parts: (1) a limit, in answer to the question "How much?", and (2) an answer to the question "How often?" The question "How often?" is referred to in statistical terms as a confidence level. Basically, a 95 percent confidence level means that 95 times out of a 100, just by chance, our answer will fall within prescribed limits. Common sense tells us that errors in sampling due to sample size will be more likely to be closer to zero than to any extreme value. Furthermore, if we have taken our sample properly, the errors are just as likely to be negative as positive; that is, the distribution of these errors will be symmetrical. For sample sizes with which we commonly deal in work sampling, and for the proportions, p, with which we commonly deal, the distribution of errors in work sampling can be assumed to be distributed in the form of a normal curve.

The normal curve is, generally speaking, a well-understood concept. We shall assume that the reader is familiar in at least general terms with the normal curve. We can use this knowledge directly and usefully to complete our statement of the reliability of a sample. If we consider the total area under the normal curve to equal 1.0, we are saying that the probabilities of any error of any particular size add up to 1.0, or certainty. It is characteristic of the normal curve that we can relate distances from the mean or center point along the abscissa in terms of σ_p to areas under the curve. Figure 14-16 shows this for even-numbered distances of σ_p. In work sampling it is quite common to use a "2σ limit" or a "95 percent confidence limit." What this means is that 95.45 percent of the random errors of estimate about the true value will fall within + and − $2\sigma_p$. This gives us a very explicit statement of possible error. So it is common practice, to give a consistent statement of the reliability of the sample, to say that the true value, p', will lie within + and − $2\sigma_p$ of the sample p 95.45 percent of the time. We have now specified "how much," which is $2\sigma_p$, and "how often?", which is 95.45 percent. We usually drop the 0.45 percent and refer to the 2σ level as the 95 percent "level of confidence."

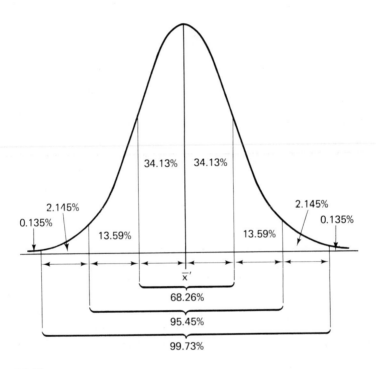

Figure 14-16.

We may be appearing to take a lot of trouble to specify the degree of reliability of our sample. It is quite true that most management quantitative measures are not subjected to such a rigorous examination. But it has been the author's experience that this may be the sticky point for many people who would benefit from the use of work sampling, but who dismiss it as just a "work sample" or "too complicated." We should always remember that the most likely value of p' (the true value) is in fact the percentage p which we have from our sample. Also, those with special interests may argue, "You, yourself, agree that your sampling proportion p may be as much as 5 percent low." The answer to this is "Yes, it may be 5 percent low, but it also could be 5 percent high," and again the best estimate still is the value of p that was arrived at by the sample. One more comment is necessary: although the 95 percent confidence level is by far the most common, other confidence levels sometimes are used. We should particularly note that the 3σ level (or 99.73 percent confidence level) is used in quality-control procedures for a manufactured product. But work sampling and indeed all work measurement represents a different situation, and the 2σ level has seemed to be more appropriate. It is suggested that this be made the prescribed level of confidence for work sampling studies.

Let us return to our discussion of sample size for a moment. It seems to be the most sensible practice to calculate the maximum number of observations that can conveniently be taken, and to do a sample calculation to find out the size of the error possible $(2\sigma_p)$ at the 95 percent level of confidence. Then a judgment will be made as to whether this is good enough for the purposes of the study. To do the calculation, some estimate will have to be made as to the value of p. Sometimes we use the case where $p = 0.50$. In this case, the absolute value of the error, $2\sigma_p$, is greatest. Sometimes we have data from other measures, and sometimes we use the results of the trial study. However, we should note that our calculation of σ_p gives us an error in absolute terms. That is, if we make a study in which a category of interest has $p = 0.40$, and $2\sigma_p$ is equal to 0.03, this means that our 95 percent confidence level lies between 0.37 and 0.43. We do not take 3 percent of p when we calculate $2\sigma_p$ in this fashion.

So far, we have talked principally about the number of observations, N, as the determining factor in calculating sampling error. The numerator of the equation, which appeared under the square root radical, is $p(1 - p)$. This of course varies with the value of p. But the variation, which may go from 0.0 to 0.50, is less a determining factor than N, the sample size. Half the values of this radical, those from $p = 0.25$ to $p = 0.75$, lie between 0.433 and 0.500. Actually, it is common in calculating sampling error to use an expression that has as its numerator twice the square root of $p(1 - p)$, and its denominator the square root of N.

$$2\sigma_p = \frac{2\sqrt{p(1-p)}}{\sqrt{N}}$$

Tables 14-3, 14-4 and 14-5 contain these values. Using these tables, it is simple to calculate the error for any work sampling category.

Table 14-3. VALUES OF $2\sqrt{p(1-p)}$ AND $3\sqrt{p(1-p)}$ FOR VALUES OF p FROM 0.01 TO 0.99*

p	p'	Confidence Limits 95% $2\sqrt{p(1-p)}$	99% $3\sqrt{p(1-p)}$	p	p'	Confidence Limits 95% $2\sqrt{p(1-p)}$	99% $3\sqrt{p(1-p)}$
0.99	0.01	0.1990	0.2985	0.74	0.26	0.8772	1.3158
0.98	0.02	0.2800	0.4200	0.73	0.27	0.8880	1.3320
0.97	0.03	0.3412	0.5118	0.72	0.28	0.8980	1.3470
0.96	0.04	0.3920	0.5880	0.71	0.29	0.9076	1.3614
0.95	0.05	0.4358	0.6537	0.70	0.30	0.9166	1.3749
0.94	0.06	0.4834	0.7251	0.69	0.31	0.9250	1.3875
0.93	0.07	0.5102	0.7653	0.68	0.32	0.9330	1.3995
0.92	0.08	0.5426	0.8139	0.67	0.33	0.9404	1.4106
0.91	0.09	0.5724	0.8586	0.66	0.34	0.9474	1.4211
0.90	0.10	0.6000	0.9000	0.65	0.35	0.9540	1.4310
0.89	0.11	0.6258	0.9387	0.64	0.36	0.9600	1.4400
0.88	0.12	0.6500	0.9750	0.63	0.37	0.9656	1.4484
0.87	0.13	0.6726	1.0089	0.62	0.38	0.9708	1.4564
0.86	0.14	0.6940	1.0410	0.61	0.39	0.9756	1.4634
0.85	0.15	0.7142	1.0713	0.60	0.40	0.9758	1.4697
0.84	0.16	0.7332	1.0998	0.59	0.41	0.9836	1.4754
0.83	0.17	0.7512	1.1268	0.58	0.42	0.9872	1.4808
0.82	0.18	0.7684	1.1526	0.57	0.43	0.9902	1.4853
0.81	0.19	0.7846	1.1769	0.56	0.44	0.9928	1.4892
0.80	0.20	0.8000	1.2000	0.55	0.45	0.9950	1.4925
0.79	0.21	0.8146	1.2219	0.54	0.46	0.9968	1.4952
0.78	0.22	0.8286	1.2429	0.53	0.47	0.9982	1.4973
0.77	0.23	0.8416	1.2624	0.52	0.48	0.9992	1.4988
0.76	0.24	0.8542	1.2813	0.51	0.49	0.9998	1.4997
0.75	0.25	0.8660	1.2990	0.50	0.50	1.0000	1.5000

*R. E. Heiland and W. J. Richardson, *Work Sampling* (New York: McGraw-Hill, 1957), p. 232.

Table 14-4. VALUES OF n AND $\sqrt{n}$ *

n	$\sqrt{n}$	n	$\sqrt{n}$	n	$\sqrt{n}$	n	$\sqrt{n}$	n	$\sqrt{n}$	n	$\sqrt{n}$
31	5.57	71	8.43	155	12.45	355	18.84	610	24.70	1050	32.40
32	5.66	72	8.49	160	12.65	360	18.97	620	24.90	1100	33.17
33	5.74	73	8.54	165	12.85	365	19.11	630	25.10	1150	33.91
34	5.83	74	8.60	170	13.04	370	19.24	640	25.30	1200	34.64
35	5.92	75	8.66	175	13.23	375	19.37	650	25.50	1250	35.36
36	6.00	76	8.72	180	13.42	380	19.49	660	25.69	1300	36.06
37	6.08	77	8.78	185	13.60	385	19.62	670	25.88	1350	36.74
38	6.16	78	8.83	190	13.78	390	19.75	680	26.08	1400	37.42
39	6.25	79	8.88	195	13.96	395	19.88	690	26.27	1450	38.08
40	6.32	80	8.94	200	14.14	400	20.00	700	26.46	1500	38.73
41	6.40	81	9.00	205	14.32	405	20.13	710	26.65	1550	39.37
42	6.48	82	9.06	210	14.49	410	20.25	720	26.83	1600	40.00
43	6.56	83	9.11	215	14.66	415	20.37	730	27.02	1650	40.62
44	6.63	84	9.17	220	14.83	420	20.49	740	27.20	1700	41.23
45	6.71	85	9.22	225	15.00	425	20.62	750	27.39	1750	41.83
46	6.78	86	9.27	230	15.17	430	20.74	760	27.57	1800	42.43
47	6.86	87	9.33	235	15.33	435	20.86	770	27.75	1850	43.01
48	6.93	88	9.38	240	15.49	440	20.98	780	27.93	1900	43.59
49	7.00	89	9.43	245	15.65	445	21.10	790	28.11	1950	44.16
50	7.07	90	9.49	250	15.81	450	21.21	800	28.28	2000	44.72
51	7.14	91	9.54	255	15.97	455	21.33	810	28.46	2100	45.83
52	7.21	92	9.59	260	16.12	460	21.45	820	28.64	2200	46.90
53	7.28	93	9.64	265	16.28	465	21.56	830	28.81	2300	47.96
54	7.35	94	9.70	270	16.43	470	21.68	840	28.98	2400	48.99
55	7.42	95	9.74	275	16.58	475	21.80	850	29.16	2500	50.00
56	7.48	96	9.80	280	16.73	480	21.91	860	29.33	2600	50.99
57	7.55	97	9.85	285	16.88	485	22.02	870	29.50	2700	51.96
58	7.62	98	9.90	290	17.03	490	22.14	880	29.67	2800	52.92
59	7.68	99	9.95	295	17.18	495	22.25	890	29.83	2900	53.85
60	7.75	100	10.00	300	17.32	500	22.36	900	30.00	3000	54.77
61	7.81	105	10.25	305	17.46	510	22.58	910	30.17	3200	56.57
62	7.87	110	10.49	310	17.61	520	22.80	920	30.33	3400	58.31
63	7.94	115	10.72	315	17.75	530	23.02	930	30.50	3600	60.00
64	8.00	120	10.95	320	17.89	540	23.24	940	30.66	3800	61.64
65	8.06	125	11.18	325	18.03	550	23.45	950	30.82	4000	63.25
66	8.12	130	11.40	330	18.17	560	23.66	960	30.98	4200	64.81
67	8.19	135	11.62	335	18.30	570	23.88	970	31.15	4400	66.33
68	8.25	140	11.83	340	18.44	580	24.08	980	31.31	4600	67.82
69	8.31	145	12.04	345	18.57	590	24.29	990	31.46	4800	69.28
70	8.37	150	12.25	350	18.71	600	24.49	1000	31.62	5000	70.71

*Op cit., p. 233.

Table 14-5.   SAMPLE SIZES REQUIRED FOR VARIOUS LIMITS OF ERROR. 95 PERCENT CONFIDENCE LIMITS†

p	Sample Size Required for Confidence Limits at 95%					p
	±0.01	±0.02	±0.03	±0.04	±0.05	
0.01	396*	100*	44*	25*	16*	0.99
0.02	784	196*	88*	49*	32*	0.98
0.03	1.163	292	130*	73*	47*	0.97
0.04	1,535	384	171	96*	62*	0.96
0.05	1,900	475	212	119	76*	0.95
0.06	2,260	565	252	142	92*	0.94
0.07	2,604	654	290	163	102	0.93
0.08	2,945	738	328	184	118	0.92
0.09	3,278	820	364	205	131	0.91
0.10	3,600	900	400	225	144	0.90
0.11	3,918	980	435	245	157	0.89
0.12	4,224	1,055	470	264	169	0.88
0.13	4,520	1,130	504	282	181	0.87
0.14	4,820	1,210	535	302	193	0.86
0.15	5,100	1,275	568	318	205	0.85
0.16	5,380	1,350	600	337	216	0.84
0.17	5,650	1,415	628	353	226	0.83
0.18	5,900	1,475	656	369	236	0.82
0.19	6,160	1,545	685	385	246	0.81
0.20	6,410	1,605	715	400	256	0.80
0.21	6,640	1,660	740	415	266	0.79
0.22	6,870	1,720	765	430	275	0.78
0.23	7,100	1,780	790	444	284	0.77
0.24	7,300	1,830	815	456	292	0.76
0.25	7,500	1,880	835	470	300	0.75
0.26	7,690	1,925	855	481	308	0.74
0.27	7,885	1,970	875	493	316	0.73
0.28	8,065	2,015	895	504	323	0.72
0.29	8,240	2,060	915	515	330	0.71
0.30	8,400	2,100	935	526	337	0.70
0.31	8,555	2,140	950	535	343	0.69
0.32	8,705	2,175	965	545	349	0.68
0.33	8,840	2,210	985	553	354	0.67
0.34	8,975	2,245	1,000	561	360	0.66
0.35	9,100	2,275	1,010	569	365	0.65
0.36	9,220	2,305	1,025	576	369	0.64
0.37	9,325	2,330	1,035	583	373	0.63

†Op cit., p. 233.

Table 14-5. SAMPLE SIZES REQUIRED FOR VARIOUS LIMITS OF ERROR. 95 PERCENT CONFIDENCE LIMITS† (continued)

| p | Sample Size Required for Confidence Limits at 95% | | | | | p |
	±0.01	±0.02	±0.03	±0.04	±0.05	
0.38	9,425	2,355	1,045	589	377	0.62
0.39	9,515	2,380	1,055	595	381	0.61
0.40	9,600	2,400	1,065	600	384	0.60
0.41	9,675	2,420	1,075	605	387	0.59
0.42	9,745	2,435	1,085	609	390	0.58
0.43	9,805	2,450	1,090	613	392	0.57
0.44	9,855	2,465	1,095	616	395	0.56
0.45	9,900	2,475	1,100	619	397	0.55
0.46	9,935	2,485	1,105	621	398	0.54
0.47	9,965	2,490	1,110	623	399	0.53
0.48	9,985	2,495	1,110	624	400	0.52
0.49	9,995	2,500	1,115	625	400	0.51
0.50	10,000	2,500	1,115	625	400	0.50

*Since, as a rule of thumb, np should equal 5 or more, the numbers followed by * should be increased to meet this criterion. For example, for p-0.03, n should be increased from 130 to 167, so that $0.03(167) - 5$.

Examples of the Use of Tables 14-3, 14-4 and 14-5

Table 14-3 is a listing of values of $2\sqrt{p(1-p)}$ and of $3\sqrt{p(1-p)}$. Usually, we use the value of $2\sqrt{p(1-p)}$ in our calculations, since this helps us solve for $2\sigma_p$. This "two-sigma" level of confidence is common to most work measurement. The values of $3\sqrt{p(1-p)}$ are included since the "three sigma" level of confidence (or $3\sigma_p$) is commonly used in Statistical Quality Control practice in manufacturing, and we want to make this relationship and distinction clear. We will use $2\sqrt{p(1-p)}$.

Table 14-4 is simply a table of square roots. Using both of these tables it is easy to reduce the calculation of $2\sigma_p$ to a simple division. The equation for this calculation, given previously, is $2\sigma_p = \dfrac{2\sqrt{p(1-p)}}{\sqrt{n}}$.

Example: Let us consider the results of the sampling given in Table 13-2. We usually check the category with the largest percentage or proportion, since this gives the largest absolute limit of error in most cases. So if we look at the Machinists' craft, we find that the value for Category 4, "Work", is 39.1% or $p=0.391$. The number of observations is 903. If we enter Table 14-3 with a value of p of 0.39, we find a value of $2\sqrt{p(1-p)}$ equal to 0.9756. If we then enter

Table 14-4 with a value of n=900, we find the value for $\sqrt{n}$ equal to 30.00. We can then either use this value (probable practice) or interpolate between this and the value of the square root of 910; that is, 30.17. Assuming that we use 900, or 30.00, and substitute in the equation $2\sigma_p = \dfrac{2\ p(1-p)}{n}$, we have $2\sigma_p = \dfrac{0.9756}{30.00}$ or 0.03285. If we round this to 0.033, we can make the following statement:

Our best estimate of p′ (the true proportion of Category #4 in the population) is 0.39 or 39%. We are confident that 95% of similar samples will give values of p which lie between our present best estimate of p (0.39) plus $2\sigma_p$ (0.033), or 0.423 and our present best estimate of p (0.39) minus $2\sigma_p$ (0.033), or 0.357. We assume that the differences in the values of the numerator of our equation ($2\sqrt{p(1-p)}$) in this range are not of consequence.

It may seem that this is a complicated way of stating the straightforward concept of the confidence level, but it should be understood that the true value of the proportion, p′, almost never is known. If it were, we would not have to sample. A statement of confidence level and possible error, on the other hand, is inherent in all sampling. Such a statement really should be considered in describing many numerical values which are taken as "accurate," but which in fact are subject to some variation. It has been pointed out, for example, that all work measurement is a form of sampling.

Notice that the results above are expressed in absolute terms. That is, the interval of three and a fraction percent is *not* three and a fraction percent of p, but rather p plus and minus 0.033. This concept in calculating and expressing possible error has already been discussed. It seems to the author to be more sensible to deal in absolute values in most practical applications.

Use of Table 14-5

A "quick and dirty" method of getting an indication of the limits of error of a sample is the use of Table 14-5. Again, the confidence level is taken at 95%, and the limits of error are stated in absolute terms.

In our example from the previous discussion, we had p=0.39 and n-903. We arrived at a limit of error of plus and minus 0.033. If we enter Table 14-5 with p=0.39, we see that our sample size of 903 lies between ±0.03 and ±0.04. Remembering the fact that error varies inversely as the square root of the sample size, we can interpolate between the value of n for limits of ±0.03 (which is n=1,055) and the value of n for limits of ±0.04 (which is n=595) and get a reasonably accurate estimate. Of course, if we want a more precise estimate, we simply use Tables 14-3 and 14-4. Even in this case, the calculation is easy to do.

The above discussion should not be taken as an indication that limits of error and confidence level are unimportant. Any statement of the results of any sampling should have these characteristics stated quite explicitly. Rather, it is a caution that the reader should not become too entranced with the mathematics of work sampling. The best testimony to this point is a remark made to the author by L.H.C. Tippett, who originated the technique, to the effect that he almost did not publish his original definitive work because he did not see much new in the theory. He was much too modest, of course, but work sampling is really more an imaginative and powerful application of accepted sampling theory than an elegant new statistical exercise. The mathematics are straightforward and should be no obstacle to widespread use of the technique.

The opinion has been expressed that, except for work sampling studies taken to establish allowances in work measurement, the concept of taking a percent of a percent in calculating reliability does not seem to be logical. We have also said that to use such a formula and to establish a "necessary" number of observations can lead one into extravagant demands for observations. For example, if we were to insist that $2\sigma_p$ be equal to, say, 5 percent of p, regardless of the magnitude of p, we would find ourselves requiring 25,921 observations where p is equal to 0.06 and 1,600 observations where p is equal to 0.50. These numbers were obtained by taking the formula that $2\sigma_p$ is equal to twice the square root of $p(1-p)$ over the square root of N. If we set $2\sigma_p$ equal to (0.05 $\times$ 0.06) and solved, we would find that 25,921 observations would be "needed." On the other hand, when p is equal to 0.50, we need 1,600 observations to meet the same requirement. Adherence to such a rule simply does not make sense.

Essentially, to apply such a blanket rule indiscriminately means that we are disregarding the practical restrictions on our ability to take observations. Any rule that applies to all groups and to all values of p is hard to justify. Use of such rules in stopwatch time studies applied to element readings makes sense. But in that case we are dealing with a situation which is supposed to be repetitive, where the sample sizes are smaller, and where we are measuring a variable, time. Furthermore, the pattern of element observations in stopwatch time studies can reveal the presence or absence of a standardized method, and tell us something about the consistency of an operator. No such needs exist in work sampling; in fact, we know we are dealing with a nonrepetitive situation and that we are sampling for attributes. So there is very little transfer in mathematics from one technique to the other.

The reliability limits about which our discussion has centered are useful in the case of an initial sample, because they completely describe the variations in that particular proportion, p. When we take successive studies, usually a few months later, to determine the presence or absence of change, they are equally useful. The theory of sampling holds that we can calculate the standard error of the difference in p, and that if the difference between successive studies is greater than twice this error, a significant change has occurred. Discussion of all these calculations is beyond the scope of this text, but the ability to measure change is important, and it can be done between successive work sampling studies. Control charts also can be made for consecutive weeks after the original sampling. This takes the form of a standard p chart and is quite straightforward.

We will compare successive weekly samples. We always can add together the weeks and get an overall monthly sample.

This discussion by no means exhausts all aspects of sample size. But the most important concept has been introduced: a practical limit exists to the number of observations that it may be possible to make in any particular study. By "possible," we mean not only that a given number will be taken, but also that the person taking the sampling will have the time and the inclination to do the job properly. In the end, the judgment as to whether the reliability of the sample is good enough for the purposes of the study will have to be made by the person directing the study. There is no universal rule to cover this.

Informing Everyone Concerned with the Study

Work sampling is a form of work measurement. We take a work sampling study to analyze the activity of people or machines; we usually intend to draw inferences from this study and, hopefully, to take some positive action to improve the situation studied. One step in any form of work measurement where operators are under direct study or direct observation is of course to inform those being studied of the fact that they are going to be observed, and to give them an idea of the nature and extent of the observation. This should be done regardless of whether or not the people to be observed are members of a bargaining unit. No one likes to be observed surreptitiously, whether he be a professional or the lowest classification on the wage roll. There are a number of reasons for informing personnel: (1) It is just not acceptable personnel practice to fail to do so. The leader has only to ask himself how he would feel if he had been studied without his knowledge and then presented with results and suggestions for change as a result of this study. (2) Surreptitious and concealed studies are sometimes explicitly prohibited in labor union contracts. Where this express prohibition does not appear, it may be because such a clause was once part of the contract, but has been dropped because company policy has a similar prohibition. (3) As a practical matter, it would be difficult to conduct a work sampling study without the knowledge of those being observed, and extremely difficult to use the results.

It has been the author's experience that after the first few days of the sample study, if the observations are properly randomized, there will be no systematic alteration of work habits. There have been cases where this has not been true, but organized opposition to such studies has been quite obvious from the start. This is a most unusual condition, incidentally. A more common experience has been that the supervisor, when he acts as observer, sometimes introduces a bias into the results. But this, too, is quite rare and usually can be spotted when sampling results and an analysis of work units of output are not consistent.

A useful means of informing the people concerned is to write a memorandum describing the study, and giving as many of the details as may be thought necessary, for discussion by the supervisors with the people to be sampled. An example of such an announcement appears in Figure 14-17.

This form of written statement of the objectives and form of procedures of the study also is useful in informing all levels of management of the fact of the study. This is not only good personnel management, but also serves as an indicator of management's interest in a cost improvement program, where the work sampling is being done as part of such a program.

Developing Randomized Times for Observation

In any form of sampling, the usual procedure is to define the population, draw a sample, and then extend the inferences of the characteristics noted in the sample to the entire population. To extend these inferences thoroughly, the sample drawing must be representative of the entire population. A safeguard in sampling procedures, to help ensure this, is the randomizing of times of observation. In work sampling, we select random times for the start of each tour of observations. These times should satisfy the requirement that every minute of the time we have defined as our sampling population has an equal chance of selection as a time of observation as every other minute; furthermore, there shall be no pattern to the selection of these times, so one time of observation may be considered as independent of any other time of observation. Examples will be given of the actual procedure, but if these basic requirements of randomness are kept in mind, the selection of random times for observation is usually quite straightforward.

We have already discussed the first step in randomization of observation times. That is, we have already said that the exact work day be specified. Let us take as a simple example a work shift that starts at 8 A.M., works until noon, has an hour off for lunch, and returns to work until 5 P.M. Let us say that specific rest periods or personal time periods are not designated, but rather the employee is allowed to attend to his own personal needs whenever he feels like it. Thus we have a solid four-hour period in the morning and another in the afternoon. We then would start with the first minute that an observation is possible, 8:00 A.M. and assign it a consecutive minute number of 000. The next minute, 8:01 would be assigned the number 001 and so forth until 11:59. This covers four hours or 240 minutes of work; since we started with 000 at 8:00, our last minute of observation in the morning would be at 11:59, whose consecutive number would be 239. The employees take one hour for lunch, during which time we do not sample. Our next time of observation would be 1:00 P.M., with a consecutive number of 240. We would then continue to assign the numbers representing the minutes until we got to 4:59 P.M., which would have the consecutive number 479. We therefore will draw random numbers from 000 to 479 inclusive. These random numbers are contained in standard tables of random numbers, such as Table 14-6. Regardless of the groupings in the table, each digit is independent of

SAMPLING PROCEDURE
_____ NUCLEAR STATION
January 11, 1973

1. A work sampling study will be conducted at the construction site over the next month or so. The purpose of this study is to analyze craftsman activity, as a reflection of supervisory and planning practice. It is hoped that this study will be helpful in initiating improvements, as well as in providing a bench mark of present practice.

2. The name of any individual craftsman will not appear on any management report or summary. No time study or appraisal of effort or method will be made. Sampling will be done by individual and craft group, and reports issued by craft group. An observation of an individual will record only his activity, as described by the list of categories of activity.

3. Two crafts have been selected as typical of the work. Pipefitters and Insulators are the crafts. A sample of the craftsmen will be used. If the results are useful, the sampling will probably be extended to include other trades.

4. The samplers will be _____ and _____. Both these men are regular employees of _____ Power Company, who have been on the site for some time, and who are qualified to do the work.

5. The sampling will be done from 8:00 A.M. to 4:20 P.M., with a break for lunch, and conforms to the project agreement.

6. Prior to the start of sampling, the study will be explained to those who will be sampled. No one will be sampled without knowing that he has been included in the sampling group.

7. Records of output will be analyzed to make certain that the work sampled is typical of general work in that particular craft.

8. The results of the sampling will be a series of percentages of the total number of observations which fall in each of the categories of activity. Under no circumstances will an individual craftsman be affected by the results of this study. Indeed, after each round of observations has been summarized, it will not be possible to identify craftsmen as individuals.

9. Work Sampling is a technique which has been widely used elsewhere in maintenance and construction. We intend to use it here as a tool to help our supervisors and engineers plan and direct the work in a more efficient manner.

Figure 14-17.

any other digit in a random number table. So we would start taking these numbers three at a time, in the specified range and simply use as a starting time of
observation whichever minute of the day matched. For example, if the first
number we drew were 137, this would represent the 137th minute, or, in our
case, 10:17.

It might be helpful to the reader also to illustrate the development of
random times of observation for a work day which is not continuous, but has
rest periods and breaks during which no sampling is done. An example of such a
work day is shown in Figure 14-10. In this case, the "working" day is 6.5 hours
out of the eight hours' clock time. To the 6.5 hours, or 390 minutes, we assign
random numbers from 000 to 389 inclusive. Let us assume that we will take five
rounds of observations per day. Starting at the upper left hand corner of Table
14-6, and taking the first three digits, then moving down to the row below and
taking the first three digits, and so forth, we have:

        157
        854—out
        476—out
        132
        105
        050
        655 out
        596—out
        313

The useful numbers which remain are:

        157
        132
        105
        050
        313

Arranging these in numerical order, we have:

        050
        105
        132
        157
        313

From Figure 14-10, we define our work day in terms of consecutive
minutes, as follows:

8:05	to	9:24	minutes	000	to	079	inclusive
9:45	to	11:54	minutes	080	to	209	inclusive
1:05	to	2:54	minutes	210	to	319	inclusive
3:15	to	4:24	minutes	320	to	389	inclusive

Therefore, the first observation would be taken in a round of observations starting in the 50th work minute, or at 8:55. The entire schedule is as follows:

Random number	Time to start round
050	8:55
105	10:10
132	10:37
157	11:02
313	2:48

The next sampling day would require different numbers, so we simply continue down, as follows:

690 915—out
    635—out
    897—out
    705—out
    141
    924—out

We have exhausted this column, so we continue by shifting over to the next three-digit set and starting upwards:

690—out
599—out
121
247
973—out
946—out
105
853—out
089

From this list, we have useful numbers:

141
121
247
105
089

As long as the objectives of randomization and the need for specific definition of the work day are kept in mind, no problems should arise.

In selecting the random numbers the usual practice is simply to turn one's head, drop a pencil at random on the page of random numbers, and start to select numbers systematically. By systematically, we mean to hold to a pattern of straight line drawing. One can go up, down, diagonally, or sidewise on the table. Since the numbers are randomized, the pattern should not be. If a particular random number is selected that is not within our restrictions, we simply ignore it. For example, the number 608 is not within our restriction of 000 to 479 inclusive. So we simply would ignore 608 and go on. If we happen to have two identical random numbers that occur on the same sampling day, there are a number of alternatives to follow. We might enlist the aid of another observer, such as an industrial engineer, and simply start from two different points in the normal tour. Or we might have the observer go out on a normal tour and then double back. This situation, or a variation of it, will probably arise during the course of the study. For example, if it takes ten minutes to make a round of observations, and we find ourselves with two random times only five minutes apart, we are essentially facing the same problem. A good idea is to enlist the aid of the second sampler, if one is available. This is particularly useful where tours of observation are quite lengthy, and employees sometimes form habit patterns based on the sampler's going by.

This last observation is a reflection of one basic reason for randomization. Even though we tell the employees about the study in advance, even though we are able to make a favorable statement to them concerning their own job security, and even though the state of employee morale is satisfactory, there still may exist a residual sense of uneasiness on the part of the employees being sampled. Although they may not feel that the sample represents a threat to their own job security, they are on notice that serious consideration is being given to altering their old familiar habit patterns. In fact, the announced purpose of the study as being part of a cost improvement program is "change for the better." The employee may be comfortable in the way he has been doing things in the past. He may remember the adage that "In order to make progress there must be change, but not all changes make progress." So he may try to anticipate the rounds of observation and introduce his own bias. It is not likely that this will happen if observations are made at random times. Employees must be told about the fact that the studies are being taken, and that they are in the group being studied, but we should avoid any indications of the exact times of the rounds of observation.

We use random numbers in selecting times for another reason. If we have a pattern to our observations, not only will the employees be able to anticipate the tours of observation, but also a pattern of observations may coincide with patterns in the activity cycle of the operation being sampled. If observations are taken on the hour, for example, certain activities might occur then with greater

Table 14-6. SOME RANDOM NUMBERS

15	77	01	64	69	69	58	40	81	16	60	20	00	84	22	28	26	46	66	36	86	66	17	34	49
85	40	51	40	10	15	33	94	11	65	57	62	94	04	99	05	57	22	71	77	99	68	12	11	14
47	69	35	90	95	16	17	45	86	29	16	70	48	02	00	59	33	93	28	58	34	32	24	34	07
13	26	87	40	20	40	81	46	08	09	74	99	16	92	99	85	19	01	23	11	74	00	79	41	69
10	55	33	20	47	54	16	86	11	16	59	34	71	55	34	03	48	17	60	13	38	71	23	91	83
05	06	67	26	77	14	85	40	52	68	60	41	94	98	18	62	20	94	03	71	60	26	45	17	92
65	50	89	18	74	42	07	50	15	69	86	97	40	25	88	14	17	73	92	07	93	11	93	45	15
59	68	53	31	48	75	47	16	49	79	69	80	76	16	60	58	53	07	04	53	66	94	94	18	13
31	31	05	36	48	75	16	00	21	11	42	44	84	46	84	83	20	49	17	12	21	93	34	61	16
91	59	46	44	45	49	25	36	12	07	25	90	89	55	25	83	47	17	23	93	99	56	14	39	16
63	59	73	21	67	80	00	25	58	25	72	06	12	86	74	54	79	70	85	88	71	58	21	98	48
89	72	47	46	94	78	56	10	65	97	84	97	42	31	49	94	15	31	13	09	45	43	03	82	81
70	51	21	03	18	50	21	99	49	73	06	99	19	24	96	39	43	10	14	12	94	08	55	54	70
14	15	99	60	44	62	72	38	18	36	63	92	61	55	93	77	66	82	10	91	81	51	67	01	47
92	46	90	39	99	64	08	00	79	27	54	96	63	40	54	34	70	27	48	18	68	59	91	83	32

Abridged from "Statistical Theory and Methodology in Science and Engineering" by K. A. Brownlee, Wiley, 1965.

or less frequency than at other times of the day. By randomizing our times of observation we also avoid any subconscious bias in the selection of times, which would amount to a surveillance of late start and early quit. This information can be gotten in other ways, if we need it. What we do is allow the concept of random numbers to operate within our carefully specified population of the work day. Periods of time in which we have a particular interest will eventually be covered, but we should not distort the nature of the technique by making it into a disciplinary tool.

A very logical explanation to the observers concerning the importance of adhering to the random times is as follows: we are particularly anxious to have a "fresh" look at a situation. If we are using supervisors as observers, the only way in which we can accomplish this is to break the supervisor's habit pattern, ask him to look at familiar activity in a different way, through the use of particular categories, and not have him take observations in his usual rounds, but rather in a pattern that breaks from the traditional. Furthermore, use of a table of random numbers opens up for discussion many of the basic principles that underlie sampling of any sort. In the course of these discussions, we should be able to increase the observer's understanding of these principles. At the very least, this would help him in any discussion of the technique that he may have with those sampled or with others.

Developing the Necessary Forms and Procedures

Work sampling is a technique for which there are many distinct uses. Many times a work sampling study is taken to serve a particular purpose, and some time later a situation occurs in which the information obtained in a study may be useful for a different purpose. In such cases, full documentation of the original study is, of course, essential. Furthermore, we usually will have learned from our experiences in the first study and want to take advantage of it in succeeding studies. And work sampling, as is true with all forms of work measurement, should be adequately documented as a matter of course. Finally, there is a very sound management principle that we should plan what we do, and reduce such planning to writing when appropriate.

The discussion that follows will deal with forms and procedures common to most work sampling studies. Some of the forms have already been referred to, and much of the procedure is discussed elsewhere. But the concept of documentation is an important one, and it is hoped that the examples will be useful to the reader. However, it must be noted at the outset that every study is unique, and that the examples shown are representative and should not simply be copied. In any application, announcements, forms, and procedures should be the product of cooperative effort of all concerned with the direction of that particular study.

An announcement of the study, in general terms, probably will be the first document to be produced. An example of such an announcement appears as Figure 14-13, (page 147). This particular announcement was prepared by the industrial engineer and other personnel of a power company that was building a large generating plant. It was directed to the personnel of the subcontractors concerned, to the union officials of the crafts concerned, and was a basis for explaining the study to the foremen and craftsmen who were to be studied. Because of the number of different audiences to which the announcement was directed, it was quite a comprehensive statement of the objectives of the study. However, almost any study should have a similar announcement, particularly because it serves as a vehicle for a uniform presentation of a legitimate management activity. This is consistent with the point already made that everyone concerned with the study should be informed of the fact of the study and of as many of the details as seem necessary. This usually includes a specification of exactly who is to be sampled.

There should also be a written definition of the work day and of the periods of the work day during which sampling will occur. Table 14-2 on page 157 is an example of a form that gives the exact definition of the work day, both that specified by the union contract and that which is in practice observed. This is important, because when the final results are presented, as in Figure 14-10 on page 144, it should be made clear that in those cases where employees are held to require travel or preparation time, this fact has been recognized. In any event, an explicit statement of the time population is necessary for the guidance of the samplers and for the extension of results.

To define both the people and the time periods that are to be sampled, and to put in perspective the degree to which results can be extended, it is also useful to identify the payroll records of the group being sampled. Figure 14-18 is an example of a time sheet which gives us absenteeism and special assignments for the group being sampled. This is also, of course, necessary to match work units of output with work activity and input.

To continue with this thought, it is necessary to identify those records of output to which we already have referred. Both payroll records and records of output usually are available now, so very little new record keeping will be involved. But it is useful to identify ahead of time those records which will be needed and to be sure that they will be available for determining unit times and also for relating the period and the people sampled to the overall operation. This, incidentally, is one reason for sampling over a calendar month. Almost all records of payroll and production are summarized in a monthly management report. We simply want to take advantage of this fact.

An observation sheet is the form upon which is recorded the code number of the category observed against the name of the employee or identification of the machine that has been observed. There is no "standard" form for this. The variables that govern the actual form are the number of people to be observed, the number of observations per day, the convenience of the observer as he moves

Daily Time Report

Job No. _1857+1870_

Date _2-13_ 19 _73_

By _M.R. SMITH_

By _____ Supervisor

Description	System	Code							Code Hours
IT-6 1-B 586 L1	B	1	0	0	0	4	0	1 6	4
IT-6 1-A 581 L12	B	1	0	0	0	4	0	1 6	10
1-D 64 - A10	B6	1	0	0	0	0	2	0 2	2
IT-6 STV CIYPS	$\frac{T1}{1}$	1	0	0	1	1	6	5 0	16
AFF 1 A TED S-1-2 6T-Y	$\frac{675}{1}$	1	0	0	2	8	1	9 5	40
A1 D-AL - S12 L12	AC	1	0	0	0	3	0	2 1	4
IT-6 49' L1226	TS	1	0	0	0	1	6	1 6	40
CR-1 IT-6	$\frac{T1}{1}$	1	0	0	1	1	6	1 6	4
									20

Badge No.	Time In	Out	Hours	Rate	C1	C2	C3	C4	C5	C6	C7	C8
*7			8	F								
20			8	M	2	5	1					
23			8		2	5	1					
36			8					8				
38			8					8				
39			8						8			
40			8						8			
37			8						8			
.26			8						8			
*9			8	F								
16			8	M						4	4	
32			8	M					8			
19	ABSENT ————————————————→											
24			8	M							8	
22			8								4	4
29			8								8	
34			8								8	
35			8								8	

Total Hours →

Figure 14-18.

Table 14-7.   WEEKLY ACTIVITY COMPARISON OF PIPEFITTER CREWS 2-2-73 THRU 2-8-73

Activity	Containment		Aux. Bldg.		Turbine Bldg. Pipe Fitters		Turbine Bldg. Instrument Fitters		Average for All Crews		
	#	%	#	%	#	%	#	%	#	%	
1.	50/168	29.8	43/114	42.1	25/114	21.9	46/150	30.7	169/546	30.9	Work–Erection and Fabrication
2.	0	0	0	0	15/114	13.2	0	0	15/546	2.7	Work–Dismantle
3.	5/168	3.0	4/114	3.5	12/114	10.5	11/150	7.3	32/546	5.8	Preparation or Clean-Up
4.	16/168	9.5	2/114	1.8	4/114	3.5	0	0	22/546	4.0	Idle
5.	41/168	24.4	32/114	28.1	35/114	30.7	11/150	7.3	119/546	21.8	Wait
6.	5/168	3.0	5/114	4.4	4/114	3.5	2/150	1.3	16/546	2.9	Travel
7.	12/168	7.1	4/114	3.5	3/114	2.6	9/150	6.0	28/546	5.1	Assignment of Work or Planning
8.	15/168	8.9	1/114	0.9	15/114	13.3	59/150	39.4	95/546	16.5	No Contact
9.	24/168	14.3	18/114	15.8	1/114	0.9	12/150	8.0	55/546	10.1	Personal

Total No. Observations = 546

Table 14-7. (Continued.)

Activity	Containment		Aux. Bldg.		Turbine Bldg. Pipe Fitters		Turbine Bldg. Instrument Fitters		Average for All Crews	
	#	%	#	%	#	%	#	%	#	%
Foremen										
At Work Area	7/15	46.7	6/15	40.0	12/15	80.0	10/15	66.7	35/60	58.3
No Contact	8/15	53.3	9/15	60.0	3/15	20.0	5/15	33.3	25/60	41.7

Total No. **Observations = 60**

Date:
By:

187

Table 14-8.   WEEKLY ACTIVITY SUMMARY
FOR CONTAINMENT CONDUIT & CABLE TRAY INSTALLATION
4-30-73 THRU 5-4-73

Craftsmen

Activity	4-30-73	5-1-73	5-2-73	5-3-73	5-4-73	Total	Activity Percentage
1.	9	10	11	7	10	47	47/111 x 100% = 42.3%
2.	0	0	0	0	0	0	0 = 0%
3.	6	2	2	3	2	15	15/111 x 100% = 13.5%
4.	0	0	3	0	1	4	4/111 x 100% = 3.6%
5.	5	4	3	4	3	19	19/111 x 100% = 17.1%
6.	0	3	0	0	2	5	5/111 x 100% = 4.5%
7.	1	1	3	1	2	8	8/111 x 100% = 7.2%
8.	0	2	0	0	0	2	2/111 x 100% = 1.8%
9.	0	2	2	3	4	11	11/111 x 100% = 9.9%

Total Number of Observations = 111

Foremen

	4-30-73	5-1-73	5-2-73	5-3-73	5-4-73	Total	Activity Percentage
At Work Area	1	0	1	1	1	4	4/15 x 100% = 26.7%
No Contact	2	3	2	2	2	11	11/15 x 100% = 73.3%

Total Number of Observations = 15

Date:
By:

around, the procedures used to transcribe or summarize the results, and in general the judgment of the person supervising the study. Sometimes the observations are made on a mark-sense or Porta-Punch card. Several examples have been given throughout this book, of which Figure 13-1 (page 128) is typical.

Summary sheets that present the results of studies also can take many forms. (Tables 14-7 and 14-8 are examples of summary sheets.) One requirement for such summaries, however, is usually that the number of observations in the sample be recorded. This gives the reader a more complete picture of the reliability of the study. More will be said about this in a later discussion of presentation of results.

In addition to these documents, it is well to reduce to writing any particular procedures that seem to be unusual. In the training of the observers, the ordinary procedures for taking observations and conducting the study should be made a matter of record. However, as the trial study is made, or even in discussions prior to this, some special arrangement or departure from the routine may be deemed advisable. By all means, these should be recorded.

Conducting a Trial Study over a Two- or Three-Day Period

One feature of good management is foresight. Everyone hates to be surprised unpleasantly. And when management is committing itself to the use of any form of work measurement, it simply makes good sense to "test the water" before one jumps in. It therefore is suggested that a trial study of two or three days' duration be made before the full-scale work sampling study is started.

Specifically, the reasons for such a trial study include the following:

1 / The observers will benefit from such a "dry run." They will have the opportunity to put into practice all the plans and procedures relating to the use of random times, random routes of observation, and just the fact of making the tours of observation part of their daily schedule. We should also remember that many of the observers will be first-line supervisors, for whom this may be the very first experience in work measurement.

2 / The people who are being observed also should have the opportunity to see just what is involved in a work sampling study. Usually, this is reassuring to them. At the very least, they will start to become accustomed to the process. At worst, any particular aspect of the study that is disturbing to them will become evident.

3 / A trial study offers the opportunity to test the adequacy of the categories of activity. It is not unusual for these categories to be modified slightly as the result of a trial study.

4 / Even though the sample size probably will be relatively small, and there may be systematic errors, as the observers are in the process of learning the procedure, it still should be possible to gain some insight into the order of magnitude of the proportion of observations distributed among categories.

5 / A trial study also offers the opportunity to test the measures of output upon which decisions have been made. Some of these may prove to be difficult to obtain, and others may prove to be less meaningful than was originally supposed. Although most of this can be foreseen, there is no substitute for an actual test.

The two- or three-day trial study probably should be made about a week before the regular study is about to begin. This will allow at least some opportunity to review the results, discuss problems, make changes where necessary, and see that everyone concerned with any changes "gets the word."

Taking and Recording of Observations and Presentation of Results

If the planning and preparation work has been thorough, the actual taking of observations should not present any particular problems. Observers should follow randomized times and routes, and should be precise in categorizing observed activity. The person directing the study should encourage observers to discuss with him any variation from the planned pattern. In most cases, the judgment of the observer is sound. He usually can be counted on to make sensible decisions in relating a particularly unusual situation to the general rules given to him. However, systematic review and discussion are helpful, since this maintains the observer's interest and avoids poor practice in taking the observations. It should be stressed continuously that, by structuring observations in terms of categories, we are really forcing the observer to step outside the usual habit pattern and to make an analysis in terms of the particular objectives of the study. Many of the details of taking observations have already been covered. It has been the author's experience that a learning curve exists in the taking of observations, and that what at first seems to be a burden very quickly resolves itself into a manageable part of the observer's workload. Particularly in the case of cost improvement programs, when the observer is quite likely to be a supervisor, there is a strong expectation that the observer will develop a positive attitude toward the usefulness of the information that he is gathering.

We have already made a brief reference to the possibility that an observer, particularly a supervisor, may introduce some bias into his observations. This is extremely unusual, but it does happen. Such bias usually is a defense mechanism, growing out of the supervisor's uncertainty or mistrust of the program. If we make it clear that the first study is analytical in nature and serves as a bench mark for future improvement, we usually can dispel many doubts. At the same time, if we keep good records of output, it will be quite obvious that not much is to be gained by introducing bias.

Table 14-9 is an example of a study in which there was in fact some bias on the part of the observer. Specifically, this was a study of a maintenance operation. The electrical foreman from the outset had expressed strong opposition to the work sampling study, and the trial study showed a pattern different from those of crafts doing work in the same buildings and subject to the same system of planning and control. As a general statement, electricians, pipe fitters, and machinists tend to have similar patterns of activity in maintaining electro-mechanical equipment. In this case, the work activity was recorded as being extremely high in the electrical craft, and travel and waiting were much lower. In addition, there were absolutely no observations of personal time in the electrical craft. Furthermore, when sample job orders were estimated, it was clear that no outstanding performance on the part of the electricians was involved. (This last was just a rough check.) When this was discussed with the electrical foreman, he stated that he would not "cut his own throat," and reiterated his negative attitude toward the study. The solution in this case was quite simple; the foreman was simply taken at his own word and given an initial schedule of work based on 56.7 percent work activity. Of course, he could not meet the demand, and was finally replaced as foreman at his own request. This is not to say that taken by itself 56.7 percent work is an unrealistic figure. But taken in context, it was an anomaly. When several other supervisors were willing to give unbiased observations, it seemed simply a matter of fairness to them not to permit such lack of cooperation.

There are many ways in which the results of a work sampling study can be presented. Examples have been given throughout this discussion. But the most important rule to follow is that the results should be reported in terms of the original objectives of the study. That is, if the original objective was simply to obtain a basic analysis of the activity of a particular group, this should be stated as the proportions of activity are listed. Since the technique is relatively straight-

Table 14-9.   WORK SAMPLING, CRAFT

	Machine Repair	Sheet Metal	Welders	Machine Shop	Pipe Fitters	Electricians
1. Travel	21.1	27.5	20.4	8.1	30.3	12.1
2. Plan	3.2	4.1	2.0	.5	5.8	7.1
3. Prep.	.6	—	—	.5	2.7	4.6
4. Work	28.2	34.0	17.9	66.0	33.8	56.7
5. Wait	22.8	18.2	39.8	13.2	21.6	11.4
6. Personal	6.7	6.2	9.2	4.3	3.5	—
7. No Contact	17.6	9.7	10.6	7.6	2.3	8.1
Total	1077	291	403	211	1389	875

Observations

forward, the knowledgeable manager who looks at the results will be able to draw some fairly sensible conclusions. The essential point is that the study be taken in a careful and conscientious manner, so that all may have confidence in the results themselves.

Summary

The preceding discussion is intended to serve as a guide to the conduct as a work sampling study and as an explanation of the technique, for whatever purpose. Almost all the material could be elaborated upon at some length. However, it has been the author's experience that in most cases the director of the study can make fairly rational decisions to meet particular local problems. Work sampling is included in this book primarily because of its value in making a first analysis in areas of indirect and service activity. It is here that the features of supervisory involvement, direct observation, consideration of work units of output, and the ability to state both results and the confidence we may have in these results seem to be most valuable.

EXAMPLE OF A WORK SAMPLING STUDY

The previous discussion of work sampling has been intended to be a basic "how-to" approach. We shall finish the discussion by giving an example. Examples can be valuable in that they show direct application of the rules and procedures discussed in general terms. There also is a danger in the presentation of an example: the reader may attempt to make his own study fit exactly what was done in the example. Every work situation is different. Variables include people, organization, previous history, management goals, employee attitudes concerning their job security, availability of industrial engineering personnel, and many other factors. So if the reader will accept the notion that his own situation must differ from that described in the example, he should find it useful.

The company under discussion was instituting a "profit improvement program." Impetus for the program came from top management, and they chose the company controller to direct the program. The objective of the program was, very simply, to improve the earnings of each division. (In one division, which did a great deal of work for governmental agencies, the name was changed to "cost improvement program" at the suggestion of the government contracting officer.) Responsibility for results lay with the division manager. Arrangements for training in work sampling, work simplification, and value analysis were the responsibility of the personnel department. The first steps in the program were essentially those outlined in Part One. That is, each supervisor made a budget analysis of input in terms of dollars or materials that represented dollars. Next, each supervisor made an analysis of his units of output. The factors that governed the selection of the people to be studied were the quality of supervision and the availability of staff support in the form of industrial engineering personnel. Generally speaking, each industrial engineer was assigned two supervisors for whom he would provide support. Obviously, the supervisors had been selected as the samplers.

Since the supervisors were not skilled in work sampling, and the idea of the formal improvement program was in itself a new one, the program was first given publicity throughout the company, and training sessions were organized for the supervisors. A training program is an excellent way to promote the program, and of course is essential in introducing the specific techniques involved in the program. In this case, as in many others, half the supervisors from a particular plant attended a morning session, and half an afternoon session. Sometimes included in the program are the supervisors who will be making their work sampling studies in the second round of studies. In this case, all the supervisors who were trained were going to make a work sampling study as an extension of the training. That is, when they defined categories, developed random times, and took the trial studies, these activities were not only part of the course of instruction but also the first. step of an actual study. An outline of this particular course appears as Figure 14-19. It should be noted here that the supervisors being trained were those whose budgets revealed heavy expenditures for personnel and equipment. Some supervisors, whose budgets showed heavy expenditure for material, were trained in the technique of value analysis. They were not included in this course.

Each supervisor brought in a suggested list of categories of activities. These were discussed in the training session. In the interest of space, these categories are shown as part of the summaries of results.

It was decided to sample over a particular calendar month. A three-day study was conducted in the last week of the previous month. The study was announced by each supervisor to his own people in a manner that he chose. The industrial engineer served as a resource, not only to elaborate upon points of technique, but also to answer any questions about the program. It was understood that the industrial engineer should enter the discussion only as a last resort. The hope always is that the supervisor himself will understand both the work sampling technique and the cost improvement program.

The results of the various studies are shown in Table 14-10. Simple observation forms were used and times randomized in the manner previously discussed. At the end of the month, an analysis was made of the various studies, and objectives were set for each supervisor. After six months, another study was to be taken to check on progress. No general statements can reasonably be made about so wide a variety of activities. But as a specific statement, if the pattern showed what seemed to be excessive "waiting," "personal," and "no contact," effort to improve planning and scheduling seemed to be in order. On the other hand, in those production areas where the people were working a substantial part of the time, a further short course in methods improvement was given, and the industrial engineer and the foreman worked together toward improvement. The important thing was that appropriate action was taken to improve each department, but that the action itself varied according to the department.

In this case, work sampling established a bench mark against which to measure improvement, got the supervisors involved, established a better working relationship between the industrial engineer and the supervisor, established an input-output relationship, and indicated the direction in which improvement should proceed. These objectives seem quite universal.

PROFIT IMPROVEMENT PROGRAM
FOR WORK SAMPLING

**Syscoms Division, General Products Division—Tool Manufacturing
from Selinsgrove, Fourth Street, and others**

Leader:	Prof. W. J. Richardson
Place:	AMP Incorporated Training Room
	Kline Village
	Harrisburg, Pa.
Time:	Morning Sessions —8:00 to 11:30 a.m.
	Afternoon Sessions—1:00 to 4:30 p.m.

Tuesday, September 4   —Instruction, Discussion and Questions on Work Sampling Technique

Wednesday, September 26   —Follow-up on Session I—Work on Projects, Clear up Technique Problems and Questions. (Each Trainee will have a chance to explore.)

Morning Group
Friday, October 12
    1:00 to 4:30 p.m.

Afternoon Group
Friday, October 19
    1:00 to 4:30 p.m.

Continue Session II Schedule and sum up on this technique; Also, on where we hope to go in the Profit Improvement Program.

Figure 14-19.

Table 14-10.   RESULTS OF WORK SAMPLING STUDIES, PROFIT IMPROVEMENT PROGRAM

Stores and Receiving Dep't.

Category	Percent
Handling material	33.5
Stocking material	2.9
Walking	3.7
Making up batches	14.3
Making up samples	3.7
Writing receipts	15.6
Talking	13.6
No contact	12.7
	100.0

Based on 424 observations

Compression Molding Dep't.

Category	Percent
Load and unload	56.1
Wait for cure	6.7
Work at bench	20.6
No contact	1.7
Talking	2.2
Walking	2.1
Personal	1.0
Wait set-up	.4
Set-up	6.2
Pill machine	3.0
	100.0

Based on 1,051 observations

Fabricating Dep't.

Category	Percent
Layout	4.3
Shearing	3.0
Punching	5.4
Drilling	1.0
Notching	2.6
Bending	4.7
Welding (spot)	5.5
Set-up machine	3.0
Soldering	30.0
Degreasing	8.0
Dessication	4.2
Securing stock	1.8
Tin-plating	6.6
Walking	6.8
Tumbling	.9
Paperwork	3.0
Personal, no contact	3.0
Other	6.2
	100.0

Based on 1,705 observations

Sales Service

Category	Percent
Telephone	5.0
Typing	19.0
Office equipment	2.0
Away from desk	8.0
Using files	5.0
Dictation	0
Reviewing documents	50.0
Discussion	9.0
Other	2.0
	100.0

Based on 2,122 observations

IV

SHORT INTERVAL SCHEDULING

Short interval scheduling is a technique used to control work activity of an irregular nature. The basic procedure is simply that the supervisor (or some other person) systematically reviews the output of a worker or group of workers, assigns new work to be done, and makes certain that there is an adequate backlog of work to be done. By "systematically" is meant once an hour, once every two hours, or even once every half-hour. Since the review is done periodically in short intervals of time, the name short interval scheduling has been applied to the technique.

Short interval scheduling is simple in concept and simple to apply. Its effectiveness depends upon the faithful adherence to schedule by the supervisors and to a lesser extent upon the availability of realistic estimates of the times necessary to do the work projected for each time interval. It should be emphasized that short interval scheduling is an *aid* to supervision. It is not a substitute for weak supervision and, as a matter of fact, requires strong support by management if it is to succeed at all. There are, however, many situations in which the application of short interval scheduling can result in substantial improvement in operations. Such improvement usually is an attractive enough goal to give management at all levels the incentive to "do it right."

15

Method of Short
Interval Scheduling

Short interval scheduling is not a work measurement technique, but it is closely allied with the philosophy of work measurement; it is usually applied in areas where the determinations of standard times are difficult. Finally, it is applied by each individual supervisor in a manner that makes the application unique to his operation and under his control. In other words, instead of having the scheduling done centrally, we put this responsibility closer to the point at which the money is being spent and the actual work units are being produced. Forecasting and control are exercised over so short a time interval that the technique is more a tool for getting out the work than it is a means of scheduling in the classical sense.

Short interval scheduling usually proceeds through the following steps:

1 / Selecting the area of application.
2 / Defining the work units of output.
3 / Arriving at a time estimate for the work units to be done.
4 / Deciding on the personnel who will be given various responsibilities.
5 / Agreeing upon control points and groupings of employees to be scheduled.
6 / Specifying the length of scheduling interval and objectives in backlog control.

7 / Designing forms and specifying procedures.
8 / Informing the people concerned.
9 / Specifying management control procedures.
10 / Performing short interval scheduling.

SELECTING THE AREA OF APPLICATION

Short interval scheduling is used where the work is not repetitive in nature, and we cannot predict exactly what will have to be done and how long it will take. If the work is repetitive, or if the work consists of elements that can be predicted fairly well, we simply set up a schedule in the most economical manner, dispatch job orders to the operators, and depend upon our reporting system to give the supervisor the information he needs to control his shop. Whether the schedule is produced by a high-speed printer or put on a Sched-U-Graph, or whether a simple checkoff of a Gantt chart is used, the principles and procedures for scheduling are well known and are being applied widely and successfully. Short interval scheduling is *not* intended to supplant the techniques now used in situations where the work is predictable and where good standards exist. Rather we can look for gains through the use of short interval scheduling in areas such as maintenance, clerical service, and other irregular activity. The general rule is that wherever a great deal of discretion exists at the level of the first-line supervisor and the operator, short interval scheduling may offer real advantages.

The specific undesirable condition that short interval scheduling may correct arises from a very human inclination to do the minimum and not to persist in overcoming minor difficulties and work interruptions. It is not that people are lazy, and short interval scheduling does not attempt to make people work at a faster pace. Instead, the objective is to exercise closer control over the loss of production caused by lack of work, nonstandard parts, lack of an inspector, and lack of a prompt decision in any area. The typical employee will "stretch" his work if he thinks that his next job may be onerous. And if he is not working under standards, he may not make much effort to be sure that his job runs smoothly. Although short interval scheduling cannot prevent "lost motion," it can bring to the supervisor's attention the causes for ineffective operation. When short interval scheduling is used, there is less likelihood that two mechanics will wait half a day for a welder, or that a clerk will take from one till five o'clock to do two hours' work.

In short interval scheduling we are simply providing a means for systematic review of progress. The typical supervisor is not afraid to make the ad hoc decisions that are so necessary in the running of a department. But if the matters are not brought to him promptly for decision, delays are inevitable. One advantage of standard times is that their use relieves the supervisor of some of the problems of work allocation. Where the work is not easily measured, however, he must pay more attention to this allocation. Short interval scheduling helps him to "force" work through his area. But it must be repeated that short interval scheduling is not a substitute for supervision; its use requires supervisory involvement and management's active interest and support.

DEFINING THE WORK UNITS OF OUTPUT

Scheduling is basically a process of allocating resources to perform given tasks in a dimension of time. In areas for which short interval scheduling is appropriate, it is characteristic that the "given task" or work unit is not regular and predictable. This does not mean, however, that it is impractical to predict and control. Rather we must use broader classifications and accept the fact that two work units which are considered to be similar in reporting may, in fact, require quite different times to accomplish. We allow for this when we schedule, and are protected by the supervisor's close attention. In addition, systematic review will bring to light those exceptional work units which should be investigated or made more standard.

The basic requirements for defining work units for short interval scheduling are the same as for stopwatch time study, for predetermined human work time systems, or for the development of standard data anywhere. In short interval scheduling we accept the fact that work units may have more variability, but we still ask the basic question: "What is the output of this work system?" In other words, what do we get in return for the input of human, material, and equipment resources? In a department store it might be the number of sales made, inquiries handled, or payment media filed. In a toolroom it might be the number of unique operations performed, these to be specified as the work goes along. Table 11-1 in Chapter 11 (see page 90) is an example of the work-unit approach. The important thing is to put the emphasis on an input-output relationship. In all too many cases, this concept is ignored, and emphasis is put on the traditional manning patterns. It is common to hear that "we've always had ten men here"; the implication is that the operation is well run and, anyway, is too irregular to control. Both notions are usually open to challenge; but to do so, it is necessary to have meaningful facts. Insistence that work units of output be identified and classified is fundamental to short interval scheduling, but it also is fundamental to any sort of measurement and control.

For short interval scheduling, work units should be selected with the following characteristics:

1 / Easy to count.
2 / Consistent with existing information systems.
3 / Backed by historical data, if possible.
4 / Key units, in that other work can be associated with them.
5 / Account for most of the input of work for the person or group.

The basic principle in this selection is, of course, that we want the input-output relationship of hours per work unit to be as meaningful as possible. We are using work units as a measure of what we get in return for the resources we allocate to a supervisor. The work unit is the foundation of cost accounting. By putting the emphasis where it belongs—on the output—we are in a much better position to control the cost. Wide variations in cost per unit are symptomatic of the need for standardization, or methods work.

The fundamental error which we put in perspective by concentration on the work unit is that of allowing blanket amounts of extra time to take care of the "unusual" cases. This is a common practice, because supervisors can make things run smoother if they have a "cushion." Such practices have the added disadvantage that the genuinely troublesome interruptions to production are harder to identify and correct. Attention to cost per unit in the short run will make clear just what the problems are.

ARRIVING AT A TIME ESTIMATE FOR THE WORK UNITS TO BE DONE

Scheduling of any kind implies that we have an idea of the time required to do the work units scheduled. Since short interval scheduling usually is done in connection with irregular work, we must use less precise standards, because the work to be done is not entirely repetitive and cannot be predicted with accuracy. The saving characteristic of short interval scheduling is that we have in the supervisor a person in a position to make on-the-spot judgments of the difficulty of the work unit. For example, in clerical work he can riffle through a pack of papers and make a rough estimate of whether a "usual" time can be applied to that pack. Or a design supervisor can draw on his experience in assigning times to the development of a particular mechanism or drawing. In both cases, the supervisor is considering a particular work unit, at hand, and on which he can form a judgment. Since the job will be reviewed in an hour or two, there is much less chance of being hurt by an error in estimation. This is also fair to the employee, because if he encounters unexpected difficulty, he knows that he will be able to discuss this with his supervisor within a very short time. Indeed, he will be expected to do this. Thus, the basic difficulty of prediction is minimized because the decisions are made much closer to the problem.

The degree of precision of standards for short interval scheduling depends, of course, upon the character of the work. If no standards exist, even for broad classes of operations, we may start with the pooled estimate of the supervisor and the industrial engineer. We do not expect the supervisor to be skilled in work measurement, but we do expect him to be able to estimate time. The industrial engineer's job is to provide the measurement skill, to be sure that the proper work units are estimated, and to appraise the worth of the estimates. This last is done to find out quickly which estimates are apt to be most in error. But, as a start, estimated times may be the most logical to use.

Using Statistical Techniques

Statistical techniques such as linear programming, multiple linear regression, and curve-fitting procedures are of great help at this stage. These are discussed elsewhere. If some strong relationships exist between work units and time, these techniques may directly develop standards that are better than estimates. If, on the other hand, there is no clear relationship, the techniques will show this and perhaps indicate the areas that should be studied. Again, the industrial engineer should have the skill in application, and the supervisor should have the shop knowledge and the willingness to cooperate. Figures 15-1 and 15-2 show the use of multiple linear regression; Figure 15-1 shows a very loose situation, and then Figure 15-2 shows the improvement due to short interval scheduling. The meaning of the various parts of the computer printout is explained elsewhere, but the characteristics of reduced constant term, improved t values, and increased coefficient of correlation show how the technique of multiple regression can analyze a situation and provide standards for scheduling.

The work to be scheduled may have been standardized sufficiently so that some type of standard data may be used. It is customary in such cases for the industrial engineer to arrive at a time for the job. The foreman then uses these times to do his scheduling. In other cases, the initial time estimate may be made as part of the cost-estimating procedure, and be carried along with the paperwork traveling with the job. Sometimes the estimates cover several scheduling periods and must be broken down into appropriate time segments. This is common in maintenance and toolroom work.

Y	$=$	a	$+$	$(b_1$	$\times$	$X_1)$
(predicted time in hours)		(constant term)		(coefficient)		(number of work units of one type)

$+$	$(b_2$	$\times$	$X_2)$,	and so forth to X_6
	(coefficient)		(number of work units of the next type)	

Actual equation: Multiple coefficient of correlation 0.49.

$$Y = 53.44 + 0.009X_1 + 0.068X_2 + 0.051X_3 + 0.001X_4 + 0.021X_5 + 0.069X_6$$

t values: 0.69   2.44   1.92   0.17   1.29   1.89

Figure 15-1.   Prediction equation based on historical data used (with some adjustments to coefficients b_1 and b_4) in a short interval scheduling application. This prediction equation reflects a loose control situation. (See Appendix.)

Equation: Multiple coefficient of correlation 0.83.

$$Y = 24.37 + 0.076X_1 + 0.071X_2 + 0.053X_3 + 0.037X_4 + 0.019X_5 + 0.068X_6$$

t values: 1.30   4.11   1.86   2.53   3.09   1.78

Figure 15-2.   Prediction equation for same work situation as in Figure 15-1, but based on data taken during short interval scheduling. (See Appendix.)

In summary, time estimates or time standards used in short interval scheduling vary widely in degree of precision. The important characteristics are that they be easy to use and compatible with the work units and scheduling interval. But times used are usually in the nature of estimates; if the work is standardized and good standards exist, short interval scheduling probably will not be an appropriate technique to use.

DECIDING ON THE PERSONNEL WHO WILL BE GIVEN VARIOUS RESPONSIBILITIES

The responsibilities in short interval scheduling are as follows:

1 / Statement of objectives and management support.
2 / Development of work units and time estimates.
3 / Specification of clerical procedures.
4 / Assignment of work to work centers and individuals.
5 / Follow-up to be sure work is done.
6 / Corrective action or reassignment of work.
7 / Appraisal of effectiveness and management action to correct major deficiencies.

These responsibilities are largely a matter of definition; some may be combined or expanded. There is no question, however, that the first and the last are uniquely the responsibility of top management. If top management does not state the objectives it hopes to reach through short interval scheduling, the program will drift. And if top management does not take an active interest in results and does not work toward correction of deficiencies, the entire program is pointless.

The development of work units and specification of clerical procedures go together in that these generally are the responsibilities of staff personnel. The supervisor must of course be consulted, but these activities are naturally assigned to the industrial engineers and procedures group.

Supervisor Responsibilities

The three responsibilities that are the essence of short interval scheduling are those which remain: assignment of specific work, follow-up to be sure work is done, and corrective action or reassignment of work. In the opinion of the author, these responsibilities should rest with the supervisor. If they are delegated, some of the major objectives may not be met. Specifically, one major objective is to keep the work moving through the system by clearing up production difficulties as they arise. The supervisor is the natural person to do this; he has both the judgment and the authority. He should be directly involved; placing someone between him and the source of information usually does not help. If the procedures are sensible, the assignment of work and control of the backlog are not onerous. And "getting out the work" is a basic part of the art of supervision, if not the most important part. Short interval scheduling involves the supervisor directly only if he makes the periodic review himself.

The alternative to having the supervisor do the scheduling is, of course, to delegate the responsibility and have the supervisor simply review progress, or have exceptions called to his attention for corrective action. There are two general schemes for delegating these particular responsibilities. One calls for the use of a clerk-dispatcher who is a regular company employee. The disadvantage is that he will of necessity make decisions the supervisor should make, and there will be little change in the situation. That is, individual employees still will pretty much control their assignments and the supervisor will not be so involved.

The other scheme for delegating the supervisor's responsibilities is to have someone from outside the company—usually from a consultant organization—actually do the scheduling and monitor performance. It is sometimes felt that this arrangement has the advantage of giving a completely objective view, and that the supervisor will thus be forced to improve his performance. This may or may not be true. Experience has indicated that such an arrangement is costly and sometimes creates personnel problems of some consequence. In the last analysis, the supervisor must manage his own people, and no one else can do this. Circumstances vary so widely that no generalization is really valid. It might make sense to have someone outside the organization provide technical assistance in the details of procedure, but it is the author's opinion that the best person to do the supervision is the supervisor. If management feels that supervision is so weak that it cannot do this, there are other, more obvious, alternatives.

In conclusion, specific responsibilities should be reduced to writing and assigned to specific individuals. This makes possible the systematic appraisal of performance and the evaluation of individual contributions. There are many patterns of assignment, but the basic objective is to improve the supervisor's ability to meet schedules for nonrepetitive work. Keeping this in mind, it seems logical to involve the supervisor at the point where the decisions are made. He will bear the ultimate responsibility for these decisions in any case.

AGREEING UPON CONTROL POINTS AND GROUPINGS OF EMPLOYEES TO BE SCHEDULED

The mechanics of short interval scheduling obviously require that there be control points at which specific task assignments are made and progress recorded. Also, since each application will differ, it will be necessary to make decisions concerning the groupings of employees to be scheduled. These two matters are not at all arbitrary and common sense is the best guide, but decisions must be made.

Control Points

To take the steps in order, we first consider control points. We are concerned with a combination of paperwork and work units. Our objective is to identify those physical locations and records which will enable a supervisor (assuming that the supervisor does the scheduling) conveniently to appraise progress, assign new work, and control the backlog. We want to do this with a minimum of new paperwork and, at least at the outset, with little change in the method of actually doing the work.

We should not start with the notion that we will need entirely new records. The typical business will find that it already is generating data on almost everything it does and that a system of control exists. For example, in most clerical work we already have built-in controls. We do not simply receive a sales order and float it into the paperwork stream, hoping that eventually it is taken care of properly. Instead, we keep subtotals, we institute copies that indicate progress, we batch orders and assign responsibility for each batch, and we purge files of orders as they are filled. It is common in most clerical systems, therefore, to have "in" and "out" baskets, and no new records usually need be kept, except for the schedule sheet.

In service operations, such as order filling, we also find records that go over a central point. In the scheduling of such activities as maintenance and job-shop manufacturing, the usual practice is to have the supervisor tour the facility, looking at each individual or group's work to check against schedule. Here the physical location may change, but, as a general statement, these tours are of great value in keeping up to the moment the supervisor's grasp of what is going on.

Group Size

The size of the group to be scheduled depends upon the character of the work to be scheduled. Sometimes a single person is scheduled, because he is performing a unique task and his work is not dependent on others. If the work is done by coordinated effort, of course, the scheduling should be by groups. The physical arrangements sometimes govern, also.

In summary, the establishment of control points and grouping of employees should follow these general rules:

1 / Take advantage of existing records.
2 / When scheduling by work unit, involve the supervisor.
3 / Use sampling techniques as necessary.
4 / Assign groups so that individual and group responsibilities are clear-cut.

SPECIFYING THE LENGTH OF SCHEDULING INTERVAL AND OBJECTIVES IN BACKLOG CONTROL

Scheduling Interval

Not much need be said about the specification of length of the scheduling interval in short interval scheduling, other than the obvious statement that the scheduling interval depends upon (1) the time to accomplish a work unit, and (2) the physical arrangements. If the time to accomplish a work unit is short, in the order of less than five or ten minutes, the scheduling interval should be an hour or possibly two hours. Routine clerical work is an example for which an hour might be an appropriate interval. Some kinds of tool and die work, some maintenance work, and design work usually are scheduled in longer intervals; four hours is common.

Physical arrangements are a factor in choosing an interval, particularly if the supervisor is to do the scheduling. If we are dealing with an office or restricted area in a job shop, we are free to schedule in any interval. In some cases, however, when the scheduler must go to the work, we must be careful not to overburden him with travel. Maintenance work frequently falls in this category. There is a temptation to let someone other than the supervisor do the scheduling if there is travel involved. This may defeat the entire purpose, however, since it is the supervisor who must take action in any case.

In starting short interval scheduling, the best rule is to use as short an interval as possible. Then when some experience is gained, the interval may be lengthened. At first, the technique may seem a burden on the supervisor, but the obvious gains in productivity usually overcome this. As skill is developed, the demands of time to run the scheduling diminish. After all, it is probably new to everyone and the weak spots in operation need to be corrected as they are revealed. This takes time. But it is safe to say that without short interval scheduling the weak spots might not be found at all. The time spent scheduling usually is a most profitable investment.

Backlog Control

The second part of this phase of the study is the development of specific objectives in backlog control. In shop language, short interval scheduling is a sort of combination of fine loading and dispatching. We shall assume that there is some form of bulk scheduling which makes available the total workload or backlog ahead of the group. The particular backlog with which we are concerned is the work immediately ahead of each person or group to be scheduled. The basic policy should be to keep this as low as possible. This runs counter to a common philosophy which holds that there always should be backlog so that no one will "stretch" work. This philosophy has merit if the employee really has no assurance that his next job is planned and waiting. In short interval scheduling, however, the idea is to *have* the next job ready.

The basic objective in scheduling is to meet targets in delivery, using available resources in the most economic fashion. If we do have too many people in a production or service facility, we can find this out by forcing work through at a steady pace until we find that there is not enough work to keep everyone busy. This is simply another way of saying that we work toward a very low backlog. A steady pressure to keep work units moving through the shop or service facility is a very effective way of discovering overmanning if it exists. In any event, an immediate goal of zero backlog tends to draw attention very quickly to any delay in production, because when such delays occur, work units scheduled for the period will not be completed; thus any undesirable situation will be brought to the supervisor's attention.

In this discussion of backlog for short interval scheduling, we are, of course, assuming that the scheduling function of production control is still operating in its normal fashion. The short interval scheduling technique simply releases work to the shop in small manageable lots that can be controlled at the point of action and decision. "Low backlog" in other words does not mean that the production or service facility should operate in a "hand-to-mouth" atmosphere. It does mean that we retain a better control and have more systematic follow-up so that the overall performance against all schedules will be more likely to be met.

DESIGNING FORMS AND SPECIFYING PROCEDURES

This step is a routine administrative matter, but it should be done. The forms and procedures taken together should be reduced to a procedures manual and treated exactly like any other set of operating instructions. Company practices will differ as to the exact form in which procedures are expressed. As a general statement, the steps of procedure outlined at the beginning of this chapter should be included as a minimum.

The specific forms used will, of course, vary. The particular production control system provides our frame of reference for the work to be scheduled. Our forms must therefore be integrated into this system. The particular set of standards (loose, estimated, or standard data) must be taken into account. Two basic forms are used more than any others:

1 / The work sheet, used by the supervisor or group leader to report actual accomplishment of work units.

2 / The performance sheet, used by the supervisor to control backlog and do his day-by-day planning.

An example of the first form is shown in Figure 15-3, and of the second in Figure 15-4. The forms are self-explanatory for the most part. It should be noted on the work sheet that some of the outputs vary widely from scheduling period to scheduling period. This is no cause for alarm, since a group is being scheduled, and simply means that the supervisor is using his judgment in assigning work to the various members within the group.

The performance sheet is the working document used by the supervisor to control the scheduling and to summarize results. The key figures here are the backlogs. Also, any unusual fluctuations in time per work unit or discrepancies between scheduled and actual work will be revealed. The basic need here is for a convenient, clear, and simple notation. This document is for the use of the supervisor and should not be looked on as a higher-management report.

In many cases, we can use an already existing specification of work unit and simply write on this a code for the time into which it was scheduled. Maintenance work, toolroom work, design work, and similar long-term tasks are quite susceptible of this treatment. When work is on a demand basis, such as in customer-oriented service work, the scheduling is apt to be done on a much shorter interval, usually a fraction of an hour, and some sort of check-work or counter is used to record actual work units accomplished.

The basic requirements of the forms, in summary, are that they be simple and at the same time be capable of comparison to the standard accounting system. This last is essential because we must maintain the integrity of the short interval scheduling system. One problem which sometimes occurs is that operators overstate performance; if we can tie together our regular audited accounting system and these special reports, we can eliminate this. There is, however, no "standard" form; as long as the basic objectives are met, the forms design is a matter of choice.

SHORT INTERVAL SCHEDULING WORK SHEET

Activity — Machine Room
Employee _____ Machine Room _____ Period From 8/9 To 8/3

Rate and Hours	Number of Regular Orders Booked	Number of Immediate Orders Booked	Number of Priority Orders Booked	Number of Registries	Number of Back Orders	Number of Change Orders
Date 8/9 Hours 72	32 34 27 21 26 54 **194**	16 9 37 **62**	62 58 75 65 54 **314**	264 45 **309**	3 60 **63**	52 17 **69**
Date 8/10 Hours 72	42 19 45 **109**	10 40 2 **52**	101 28 42 62 **233**	45 8 176 **229**	21 25 23 155 74 **358**	56 **56**
Date 8/11 Hours 71	76 68 38 27 **209**	7 86 7 28 **128**	15 111 31 21 **178**	2 236 1 **239**	9 153 7 89 **258**	8 13 2 18 **41**
Date 8/12 Hours 55½	50 39 25 27 27 **168**	76 **76**	85 95 69 **249**	9 198 **207**	32 **32**	42 30 **72**
Date 8/13 Hours 55½	39 63 57 **159**	33 19 99 **71**	82 52 78 **222**	237 **237**	14 4 **18**	43 15 **58**
Date Hours						

Figure 15-3. Work Sheet.

AMP #2303			Week of			Function			

SHORT INTERVAL PERFORMANCE SHEET

			Monday				Tuesday				Wednesday				Thursday				Friday			
			1	2	3	4	1	2	3	4	1	2	3	4	1	2	3	4	1	2	3	4
Unit of Work		Sched.																					
		Actual																					
Est. per 2 hrs.		Backlog																					
Max.	Min.	Total																					
Unit of Work		Sched.																					
		Actual																					
Est. per 2 hrs.		Backlog																					
Max.	Min.	Total																					
Unit of Work		Sched.																					
		Actual																					
Est. per 2 hrs.		Backlog																					
Max.	Min.	Total																					
Unit of Work		Sched.																					
		Actual																					
Est. per 2 hrs.		Backlog																					
Max.	Min.	Total																					

Figure 15-4.   Performance Sheet.

211

INFORMING THE PEOPLE CONCERNED

Short interval scheduling is a form of management control. It is simply an extension of existing scheduling practice. To someone who has no experience, it might seem that its introduction is quite definitely a "management prerogative," and that there should be no need for formal justification to all employees. The fact is, however, that the objectives of short interval scheduling may seem to conflict with the objectives of the work force. Usually no incentive is involved and no demands are being made in terms of a higher work pace. The elimination of production or service delays and more uniform output of work are certainly recognized goals of management. But short interval scheduling represents a change. If the supervisor does the scheduling, he will be checking on performance more frequently than formerly. If present practices have resulted in overmanning, this will be exposed. No thinking person will fail to recognize that both supervisors and employees will have greater demands put upon them, and that introduction of short interval scheduling may upset working-habit patterns. In fact, the intent of short interval scheduling may lie exactly in this direction. If management were satisfied with the existing patterns, they probably would not start the short interval scheduling.

The most obvious purpose in informing employees and supervisors is to show that management has a clear set of objectives and is determined to carry these out through short interval scheduling. If these conditions are met, the rest of the details come under the heading of good personnel practice. Figure 15-5 is an example of a memorandum announcing a short interval scheduling application. This particular company undertook this first study as a training activity and intended to evaluate the results for possible expansion of the use of the technique. A statement similar to this is very useful. If the supervisor is to do the scheduling, he can use such a statement in talking with his people. If outsiders are to do the scheduling, more difficulties usually are experienced with respect to acceptance by supervisors and employees. But in any case there is need of a statement of purpose and responsibility.

In informing employees, we usually rely on a broad statement such as was discussed in the preceding paragraph, followed by discussion of the details as these affect the employee. Here the procedures manual is useful. The major problems do not lie in the procedures, as a rule. The basic point to be made in explanation is that a situation now exists which needs correction. This must be given in terms of specifics. Then the most favorable honest statement about the job security of the employees should be made. For example, if one objective is to eliminate overtime, this should be stated. If we hope to reduce the work force by normal attrition, we should say so. Normally, we do not really know what the end result will be, of course. But we should realize that job security is uppermost in the minds of the employees.

One advantage of using the supervisor to do the scheduling is that because he is involved and has the responsibility he is more apt to reflect management's

SHORT INTERVAL SCHEDULING

Primary Objectives

To train supervisors and systems analysts in the use of the short interval scheduling technique. In order to do this, we will use the project method of instruction. Tentatively, the Order Department has been selected as the first area of application.

We suggest that a critical evaluation be made of the results of this project. If these meet our objectives, we would expect extension of the use of short interval scheduling to other activities, as part of the Supervisory Development and Profit Improvement Programs.

Secondary Objectives

To realize the benefits of improved scheduling. Specifically, we hope to accomplish the following in the first area of application:

1 / Elimination of present overtime. This overtime is required to clean up peak loads resulting from uneven scheduling.

2 / To provide stock status reports and booking information on a current basis.

3 / To improve productivity by scheduling work in a uniform flow. At present, workload is uneven, and productivity therefore varies widely.

4 / Improved customer service by prompt acknowledgement. This should reduce the number of inquiries and speed up the handling of inquiries that are made by the customer. This last should come about because of the reduced amount of backlog in process.

Method of Instruction

The concept of short interval scheduling is straightforward and easy to grasp, but the development of standards is unique to each area and the discipline of the reporting and work allocation systems must be maintained. This requires close and continuous attention. Therefore, a well-defined, limited area should be used as the basis of study. The project method allows us to install short interval scheduling, training the supervisor and other personnel in the department, and the systems analyst responsible for the technical aspects. _____ will direct the study in that he will organize and coordinate the work to be done and provide technical assistance.

Figure 15-5.

views forcefully. If an outsider does the scheduling, a different attitude on the part of the supervisor may result. There are some advantages to using outside help in scheduling, but in the area of personnel relations the presence of outside people may create real problems. In any case, the employees who are to work under short interval scheduling are entitled to an explanation of the technique and the objectives of its use. One has only to consider the alternative to know that such an explanation is simply the use of common sense.

SPECIFYING MANAGEMENT CONTROL PROCEDURES

The basic purpose of short interval scheduling is to help the supervisor do his job. Specifically, the technique sets up a planning, scheduling, performing, follow-up sequence that we all recognize as essential to good management at all levels. Of course, we consider the supervisor as part of management, but the substance of this section deals with the second- and third-line managers who must enter actively into short interval scheduling to make certain that objectives are being met and that procedures are being followed. Thus there must be review and control procedures. These need not be complicated and above all should not require any substantial amount of paperwork. But second- and third-line supervisors must become involved in short interval scheduling, and this should be done on a systematic basis.

The reasons for the involvement of second- or third-line supervision have their roots in the problems that short interval scheduling is intended to solve. The general problems are the inherent unresponsiveness of the work systems to change and the lack of continuous pressure to keep the work flowing through the system. We must remember that the reporting mechanisms which existed before short interval scheduling probably were too broad, and that direct pressure was hard to achieve. Therefore, such reports were not regarded as working documents or were altered after the fact to reflect what happened. Short interval scheduling should work on the principle of exceptions. But if the first-line supervisor does not have his work reviewed, he will very soon lapse back into the old way, and will properly regard short interval scheduling as simply another burden that makes his job more difficult without producing anything worthwhile. Therefore, the supervisor's manager should periodically (daily, at the start of a study) go over the schedule with the supervisor and the analyst. He should find out why there were readjustments, discuss failures to meet scheduled goals, discuss the adequacy of the standards, and in general keep a weather eye on the effect of short interval scheduling on improvement of productivity. As the study progresses, this review may be done at less frequent intervals. But it must be done if we are to achieve results. By working with the first-line supervisor and the analyst from the original documents, the manager demonstrates his interest and can control the conduct of the study.

There is a positive side to such involvement of the manager. He has a broader view of the work system and is better able to correct undesirable conditions, if for no other reason than his greater authority. In one case, a report of shipments was due to be sent from the computer facility to a production control office by nine o'clock in the morning. Frequently, it had not been sent up until the afternoon. Without this report a group of clerks really were nonproductive. Before the introduction of short interval scheduling, the supervisor simply waited for the report and his people loafed.

When short interval scheduling was introduced, he became more actively interested, of course, because he simply did not have the work units of output to account for his peoples' time. He himself had been unable to exert enough influence in the computer operation to get the report to his office at the time agreed upon. But his manager was able to resolve the problem. The key notion here was that the manager *had* to resolve the problem, or he could not reasonably hold the supervisor to the schedule. Basically, the problem had existed and been recognized for some time, but until the manager became involved directly and had the facts spread out so that he could establish his case, no action was taken. It is easy enough to say that the situation could have been corrected without short interval scheduling; this is true. But the fact is that such problems exist; they may even be typical. To be solved, they must be identified, and sufficient management interest must be generated to force their solution. Short interval scheduling is a tool that helps to do these things.

One general aspect is particularly appealing to the manager: he has a structured format within which he can get a quick and comprehensive review of the operations for which he is responsible. In brief, he does not have to depend entirely upon his intuition, but has a starting point for review and a logical sequence for monitoring actual performance versus scheduled performance. Also, if he makes it a point to do this quite frequently, he will get to know his supervisors better, and morale and communication should be much improved.

The only paperwork involved in this should be a log of the major reasons for missing schedules (Table 15-1). If less time is taken than estimated, the standard should be changed. If more time is taken, either the standard should be changed or the situation corrected. If the manager pays attention to deviations from standards, he usually gets a pretty good understanding of the effectiveness of the operation. Above all, he is able to act reasonably in helping his supervisor overcome difficulties as revealed by the short interval scheduling. It is this which in the long run is one of the most important benefits of use of the technique. The heart of management control, however, remains the manager's willingness to take the time to review progress in a systematic way. This is essential.

Table 15-1.   LOG OF MISSED SCHEDULES

| | | *Schedule Miss Report* | | |
Item No.	*Department*	*Work Units*	*Reason*	*Man-hours Difference*
1	Fabrication yard	Plates moved	Magnet on crane malfunctioning	12
2	Weldment shop	Inches welded	No print	8
3	Shipping dept.	Tons loaded	Common carrier late	10

PERFORMING SHORT INTERVAL SCHEDULING

The performance of short interval scheduling is straightforward with respect to theory and procedures. The supervisor takes the work units either from a master schedule made up by production control (Fig. 15-6) or simply works on a demand delivered to him from a previous operation from a series of work orders or from telephone or immediate customer demand. By appropriate means he then develops times for the work units, aggregates them into appropriate blocks, enters them on the schedule, and at the appropriate time assigns them to his people to be accomplished during the next interval. If all goes in a routine fashion, he simply repeats the process of scheduling, and at the end of each time interval he records the actual production and assigns new work. If for any reason an operator feels that he will be unable to reach or approach the scheduled task, he is instructed to inform the supervisor immediately so that the situation can be corrected. If the supervisor has not been so informed, he expects the work to be done on schedule.

The specific steps are implicit in the working documents (Figs. 15-3 and 15-4). The short interval performance sheet (Fig. 15-4) is the supervisor's master control. On this he has all the information for scheduling. A most significant entry here is the "backlog." By using this to adjust to workload, and by working toward its reduction, instances of overmanning become evident. Also, we issue to the operators the maximum amount of work units to be done, and will not accept without explanation any performance below the minimum. This performance sheet reflects an interval of two hours. The work sheet (Fig. 15-3) simply is the individual employee's or group leader's record of actual performance. In this case, standards had been set using multiple linear regression, so the inclusion of the "total" figure in the box in the lower-right corner of each square made it easier to check the standard.

The reader can see, however, that the paperwork and procedures involved need not be complicated. Indeed, they should be simple, because the manager's review, mentioned previously, takes the place of voluminous reports. The sys-

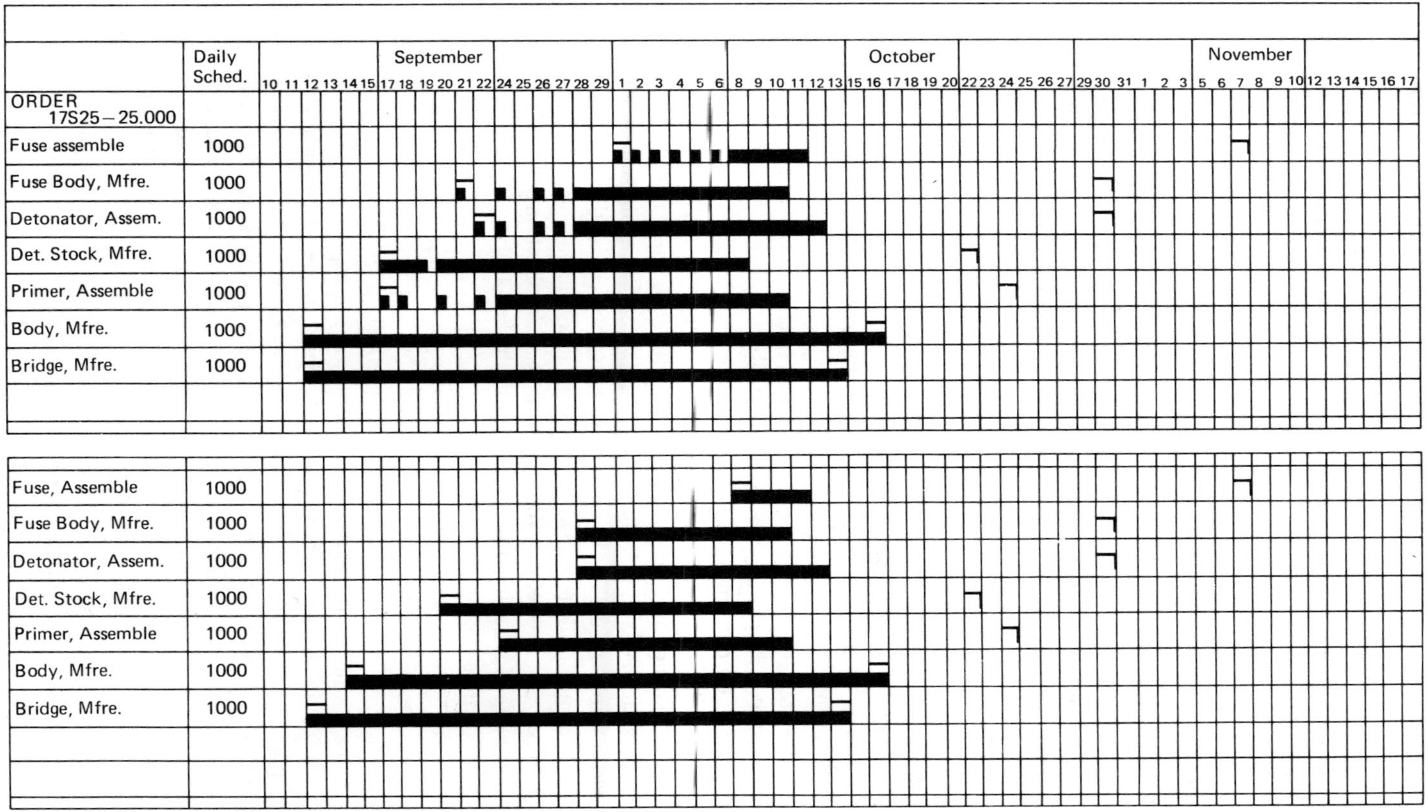

Figure 15-6. Example of a Master Schedule. Gantt progress chart for a fuse. (From G. B. Carson, Ed., Production Handbook, Ronald Press, New York.)

217

tems work involved should be complete enough to give the supervisor a series of short-range objectives for his people; the mechanics of this should be simple. The end results should be "scheduled" work units that are reasonable and "actual" results that are accurate.

Importance of the Supervisor

It is at this point that the ability of the supervisor is tested. For the supervisor must make it quite clear by his actions and attitude that he expects the work to be done, and on schedule. It is important that the standards be reasonable, of course, because these are usually the first point to be challenged. But the basic objective of the technique is to provide the supervisor with the means to get out the work. The supervisor can now furnish each employee with a reasonable short-term goal. It is then up to the supervisor to insist that this goal be met. All our planning, systems work, standards development, and training will be useless if the supervisor does not use his position and personal leadership qualities to effect the performance of his people so that they meet the scheduled output.

If we remember that by introducing short interval scheduling we are actively seeking change in the existing situation, we put in perspective the demand for good performance. It is quite likely that insisting on meeting schedule will be a departure from past practice. Also, the very supervisor who is now making the demand may not in the past have done so. The problem in the application of short interval scheduling often is one of changing the supervisor's attitude toward scheduling discipline. We must be fair to the supervisor; specifically, we must provide him with good standards, help him to overcome operational difficulties, and let everyone know that all levels of management support the short interval scheduling program. But in the last analysis it is the supervisor who is responsible for results.

It is unfortunate in a way that the technique of short interval scheduling is so named. The words are accurate, it is true, but the objective is one of supervisory action for results. The word "scheduling" implies the planning phase of management, whereas in most cases we want to improve the planning but, more importantly, the action of carrying out the plans. In the performance phase of short interval scheduling we finally get down to the essential supervisory skills that are so vital to good operation. The technique provides the supervisor with the means to direct the efforts of his people in a more effective manner. But it still is up to the supervisor to do the job. The poor supervisor traditionally has welcomed the absence of good standards and scheduling, for he then cannot be made to "look bad." But the good supervisor is perfectly willing to work with reasonable standards, because he can then demonstrate his ability and move up. The ultimate objective in all our management effort is of course profit, and better utilization of men and equipment can only improve profit.

Appendix

Measurement of Indirect Work Using Multiple Regression*

Work measurement is commonly associated with the stopwatch or with the use of predetermined time values, and has traditionally been applied to direct or repetitive work. But as the character of work has changed in recent years to include more work of an indirect nature, such as clerical, maintenance, and service activity, it has become apparent that the stopwatch and tables of predetermined time values may not be the most appropriate techniques for economical measurement. This is particularly true when the first attempt is being made to bring these indirect costs under control. In this chapter we shall discuss the use of multiple linear regression to obtain time standards for indirect work.

Multiple linear regression is a statistical technique; it may be thought of as an extension of the familiar line of least squares, or simple linear regression. Most readers have used the line of least squares technique to develop standard data or for estimating costs. Very simply, the technique enables us to associate

*Presented at the International Conference on Production Research, Birmingham, England, April 1970 by the author.

INT. J. PROD. RES., 1971, VOL. 9, No. 4, p. 481-486.

Reprints published by Taylor and Francis, Ltd., 10-14 Macklin Street, London WC2B 5NF.

an independent variable, such as the number of documents processed, with the time required to process these documents. To establish a general relationship, we must have some count or measure of the independent variable with the actual value of the dependent variable. We get these data from our experience and use them to develop a prediction equation. From then on, we can predict the value of the dependent variable (for example, time) given a value of the independent variable (for example, number of customer orders processed). The equation takes the form $y = a + bx$.

The significant facts here are that we must identify the work unit of output—what is it that we are doing or producing—and that we have some recording of the time taken to do each lot or batch. The assumption is made that there is a linear relationship between the time to accomplish work units and the number of units. The constant, a, is attributed to setup time or to some type of activity that is not directly related to the unit time. The coefficient, b, is the time to accomplish one unit of work once the job is set up, and not considering any setup or other nonproducing time. We assume that the reader will consult a standard text for any discussion of the mathematical development he may need.

The preceding discussion of simple linear regression is intended to be an introduction to the main topic, to reassure the reader that he does in fact have some background and that the problems lie in a familiar area. Multiple linear regression is an extension of the line of least squares that allows us to handle more complex situations in which many factors may be classified as independent variables and the fundamental relationship cannot be derived graphically. In particular, in many cases in indirect work we have the overall time for a group of different work units, but no indication of how much time was spent on each. This case is discussed in the second example. We also have problems in measuring work units that are nominally the same, but which have different characteristics.

EXAMPLES

Let us take as an example the use of multiple linear regression to develop time standards for cost control in the machine room of a sales order office. The machines are Flexowriters that produce punched paper tape and hard copies of sales orders. The office had been experiencing some problems in maintaining a satisfactory work throughput, and it was decided to use short interval scheduling. To do this, however, it was necessary to have some sort of time standards for scheduling. The only available historical data were records of the total hours per day and the total output. These were not "working" hours, and there was no indication of how much time had been devoted to each type of work unit. Therefore, it was expected that the predictive equation might be inaccurate, particularly so since the machine room was a problem area. However, standards were needed, and so multiple linear regression was used to develop them.

First, the variables must be identified. Time was of course the dependent variable and is referred to as Y. The independent variables are X_1, number of regular orders processed; X_2, number of immediate orders processed; X_3, number of priority orders processed; X_4, number of registries of orders done; X_5, number of back orders processed; and X_6, number of change orders processed. Our procedure is to use data based on our experience to solve for the constant, a, and the various coefficients, b_1, b_2, b_3, b_4, b_5, and b_6, in the regression equation:

$$Y = a + b_1 X_1 + b_2 X_2 + b_3 X_3 + b_4 X_4 + b_5 X_5 + b_6 X_6$$

Once the constant and the coefficients have been established, we can then predict the total time, Y, in any succeeding scheduled workload consisting of any values of X_1, X_2, X_3, X_4, X_5, and X_6. The original data are shown in Table A.

Table A.

Date	Code No.	Y	X_1	X_2	X_3	X_4	X_5	X_6
3/30	01	80	197	155	211	360	171	17
3/31	02	80	187	113	194	236	196	71
4/1	03	80	394	125	204	113	128	53
4/2	04	80	187	61	191	407	181	70
4/3	05	80	121	25	193	317	210	86
4/6	06	72	136	67	140	329	210	18
4/7	07	64	187	24	169	402	174	69
4/8	08	80	133	122	216	259	104	84
4/9	09	80	172	97	273	329	100	00
4/10	10	72	214	55	231	406	114	69
4/13	11	78	383	100	178	396	67	69
4/14	12	80	325	115	212	381	162	36
4/15	13	80	321	100	152	82	195	74
4/16	14	80	216	30	203	335	299	67
4/17	15	80	96	32	282	514	130	34
4/20	16	80	221	25	204	96	133	68
4/21	17	80	218	50	221	404	225	47
4/22	18	80	170	94	151	621	188	58
4/23	19	80	247	83	282	175	124	76
4/24	20	80	266	52	199	396	203	74
4/27	21	80	279	75	192	419	76	47
4/28	22	74	160	17	160	263	281	85
4/29	23	80	315	49	219	122	324	40
4/30	24	80	245	91	199	546	89	24
5/1	25	94	330	116	184	86	192	128

There will be no discussion here of the mathematical development of the regression equation. Practically speaking, the next step is to put these data into the form of punched cards and to use a computer library program and an electronic computer to solve for the constant and coefficients. Every commercially used computer has a multiple linear programming package, and most computer installations already have used this package for other reasons.

The output of the computer will vary with the particular library program, but almost all have the following features:

1 / *The constant and coefficients that allow us to write the prediction equation.*

In this case these are:

			Hours
a	(the constant)	=	53.44
b_1	(coefficient or multiplying factor for X_1)	=	0.009
b_2	(coefficient or multiplying factor for X_2)	=	0.068
b_3	(coefficient or multiplying factor for X_3)	=	0.051
b_4	(coefficient or multiplying factor for X_4)	=	0.001
b_5	(coefficient or multiplying factor for X_5)	=	0.021
b_6	(coefficient or multiplying factor for X_6)	=	0.069

The regression equation is

$$Y = 53.44 + 0.009X_1 + 0.068_2 + 0.051X_3 + 0.001X_4 + 0.021X_5 + 0.069X_6$$

where Y is the predicted time in hours. The value, a, of over 53 hours, when compared to the average of about 79 hours may be interpreted to mean that we do not have a really sensitive equation in terms of explaining or accounting for the influence of workload on time to accomplish the work. In other words, there is a relatively large constant value, which may be due to the presence of work units of output that are done but not included in the independent variables, or by a large amount of personal delay time, or by some other factor. It is dangerous to draw conclusions, but as a general statement this large constant reflects our impression that the processing of work through the machine room is rather haphazard. We need now only substitute for X_1, X_2, and so on, the workload in these items to get the time necessary to do the work.

2 / *A multiple coefficient of correlation squared.* This is a measure of the agreement between the predicted results (in the original data used to develop the coefficients) and the actual results. There are cases where this should be used with caution, as should any individual measure,

but it is a valuable indicator of the closeness of our prediction. In our case, this was

$$(0.49)^2 \quad \text{or} \quad 0.24$$

This indicates poor predictive value.

3 / *The t-values for each coefficient, or an indication of the reliability or significancy of each.* This is a real oversimplification, but in general, depending upon sample size, a *t*-value of over 3 indicates very great significance, of over 2, significance, and of over 1.6, probable significance. Below 1.6, *t*-values lead us to question the worth of the coefficient. Our *t*-values were as follows:

$$X_1 = 0.69 \qquad X_4 = 0.17$$
$$X_2 = 2.44 \qquad X_5 = 1.29$$
$$X_3 = 1.92 \qquad X_6 = 1.89$$

From these it can be seen that X_1, regular orders, and X_4, registries, seem to present the greatest problems, because there is more variability to our predictions of the contribution of each to the value of the dependent variable.

In summary, this first regression shows that we have a nonstandardized operation in the machine room, and that we should use the results with caution. However, we knew the situation was not good. Furthermore, we plan to use these results in conjunction with short interval scheduling to improve the situation. At least these scheduling times did not cost much, and they are the best we have.

For a second example we can use the preceding data but for a different time period. In fact, these data were taken while the short interval scheduling was being done, and there were several positive indicators that the machine room was being managed better. Therefore, the regression in this example will show a much more predictable situation, and will reflect a stronger relationship between input in hours (dependent variable) and output in work units (independent variable).

1 / *The constant and coefficients,* as in the first example, are as follows:

		Hours
a	(the constant)	= 24.37
b_1	(coefficient for X_1)	= 0.076
b_2	(coefficient for X_2)	= 0.071
b_3	(coefficient for X_3)	= 0.053
b_4	(coefficient for X_4)	= 0.037

$$b_5 \quad \text{(coefficient for } X_5) \quad = 0.019$$
$$b_6 \quad \text{(coefficient for } X_6) \quad = 0.067$$

The significant factor here is the reduction in this constant from 53 hours to 24 hours. In this case, we can take this to mean that we have reduced the amount of setup or nonproductive time.

2 / *The multiple correlation coefficient* went from $(0.49)^2$, or about 0.25, to $(0.83)^2$, or about 0.69. This indicates a much better prediction.

3 / *The t-values* improved overall; in particular, we no longer had two very weak values. The *t*-values were

$$X_1 = 1.30 \qquad X_4 = 2.53$$
$$X_2 = 4.11 \qquad X_5 = 3.09$$
$$X_3 = 1.86 \qquad X_6 = 1.78$$

In summary, therefore, we have developed a new regression equation that we are quite sure will predict the time necessary to accomplish a workload made up of the work units in the machine room. Furthermore, we know how good our prediction is and can make further analyses as changes are made.

In this appendix we wish only to acquaint the practitioner in work measurement with the concept of using multiple linear regression as a tool for work measurement. Multiple linear regression has many nuances, and interpretation of the results is not really as straightforward as it may appear. However, competent advice is widely available. The author feels that the technique has demonstrated its potential and deserves serious consideration, particularly in first attempts to get under control the increasingly important areas of indirect work.

INDEX